Liaison Functions 2021

AA009004

Held at the 2021 AIChE Spring Meeting and 17th Global Congress on Process Safety

Online
18 - 23 April 2021

ISBN: 978-1-7138-5343-5

Printed from e-media with permission by:

Curran Associates, Inc.
57 Morehouse Lane
Red Hook, NY 12571

Some format issues inherent in the e-media version may also appear in this print version.

Printed with permission by Curran Associates, Inc. (2022)

For permission requests, please contact AIChE
at the address below.

AIChE
120 Wall Street, FL 23
New York, NY 10005-4020

Phone: (800) 242-4363
Fax: (203) 775-5177

www.aiche.org

Additional copies of this publication are available from:

Curran Associates, Inc.
57 Morehouse Lane
Red Hook, NY 12571 USA
Phone: 845-758-0400
Fax: 845-758-2633
Email: curran@proceedings.com
Web: www.proceedings.com

TABLE OF CONTENTS

619468 Leading with Purpose, Heart and Mind

Abstract:

Mrs. Kadri will share her inspiration and leadership principles which have guided her through her career, from the labs to the Board meetings.

As a fellow chemical engineer, she will also share her perspective on how engineers are positioned to support human and economical development towards a safer, greener and more sustainable industry and economy.

620743 Networking in a Virtual World

Abstract:

Networking is the most honorable and valuable endeavor in which you can engage, because it is built on a spirit of generosity. At its core, networking is all about crafting win-win alliances where both parties provide value. You may think that "networking" can only take place in person, but this is a myth! In fact, most networking takes place from afar, and in some cases, the individuals may never even meet in person. In this webinar, we will discuss strategies and tactics for finding new collaborators and building mutually-beneficial partnerships with professionals across the globe (and perhaps on some exoplanets too!). We will utilize various methods to find and reach out to people, and start and nurture those critical conversations which can birth amazing collabs! Let's network!

618671 Simulation of a Moving Bed Chemical Looping System for Electricity Production from Coal Via Chemical Looping Water Splitting

Abstract:

Coal is an essential energy source for modern industry. However, being the most carbon-intensive fossil fuel, the combustion of coal for power generation contributes to a significant proportion of greenhouse gas emissions. To reduce CO_2 emissions from coal-fired power plants, various approaches can be adopted, such as the pre-combustion CO_2 capture and the post-combustion CO_2 capture methods. However, these CO_2 capture methods are usually energy intensive. For example, the energy penalty of an amine-based post-combustion CO_2 absorption system can reach up to 20-30% of the total power production from the plant.

Chemical looping technology is a versatile platform for industrial material and energy conversions, and it provides promising alternatives for CO_2 capture from coal-fired power plants. This work studies the application of chemical looping water splitting (CLWS) technology to replace the gasification unit and pre-combustion CO_2 separation units in a conventional integrated gasification combined cycle (IGCC) process for electricity production. The CLWS system consists of a reducer, an oxidizer and a combustor. In the reducer, coal reacts with metal oxides to be converted into CO_2. The reduced metal oxides react with water in the oxidizer to produce H_2. Finally, the metal oxides are fully regenerated in the combustor by reacting with air. H_2 is then combusted in a combined cycle to generate electricity. The CLWS process is simulated in ASPEN Plus, and the operating parameters such as the metal oxides composition and the system pressure are adjusted to maximize the power production. Process simulations indicate that the CLWS process can achieve an 18% improvement (relative percentage) in electrical efficiency compared to conventional IGCC systems, demonstrating the great potential of the CLWS technology for electricity generation with a higher energy efficiency and minimal CO_2 emissions.

619094 Heat Integration and Optimization of Low-Temperature Carbon Dioxide (CO2) Removal Via Distillation

Abstract:

Pinch analysis is one of the ways that the chemical process industry uses to optimize the process. In the low-temperature process industry, such as gas and liquid natural gas processes, heat integration is crucial to address the abundance of energy requirements by the process to meet the product specification. Current natural gas resources with low acid gas contaminants such as carbon dioxide (CO_2) is depleting. Hence, the exploration to monetize on a high CO_2 gas field was initiated. Low-temperature distillation for separation of bulk CO_2 to the acceptable specification was selected as one of the CO_2 capture technologies. However, the energy required for such application is relatively large. Therefore, a pinch analysis in the pilot design of the technology was conducted to evaluate the potential energy saving by employing the heat integration analysis. The Pinch Analysis technique uses a parameter called DT_{min} to establish the minimum theoretical energy requirements of the system. The process scheme was constructed using Aspen Hysys to determine the actual hot and cold utility load. Then, hot and cold streams and properties will be extracted into Aspen Energy Analyzer (AEA) to perform optimal heat exchanger network design. The pinch analysis and Heat Exchanger Network (HEN) results show 25% to 85% utility saving for hot and cold streams. Based on the existing process configuration, the calculated DT_{min} is greater than 5°C, but in this study, DT_{min} at 3°C is used to optimize the process and established a new HEN. Thus, the new HEN's retrofit design was incorporated to improve the hot and cold utilities target further.

RDR eReactor: CO2-Free Steam Cracking

Harri Johannesdahl
CEO
Coolbrook Oy

Abstract:

Coolbrook RDR (RotoDynamicReactor - a turbomachine for cracking hydrocarbons) technology is an innovative technology to provide significant reduction to CO2 emissions by electrification of steam cracking and at same time improving ethylene yields. Coolbrook has developed multiple concepts for reduction of CO2 emissions and will present estimated CO2 reductions of two concepts to replace conventional furnace section by RDR technology. In addition, current status of the pilot project will be presented. Detailed design and construction of pilot plant shall start during the spring 2021. By the end of 2021 first testing phase with air have been completed and the pilot site will be in construction phase.

CO2 reduction by RDR eReactor technology

RDR eReactor technology can be electrified with potentially zero CO2 emissions and it is flexible to accommodate fluctuations in renewable energy production. RDR eReactor can be implemented step by step at existing cracker site into the existing infrastructure. In this way both the environmental and commercial viability are maximized. Currently, it is technically and economically more viable than any other existing or announced furnace replacement technology to reduce CO2 emissions.

Coolbrook has evaluated process concepts for furnace section and calculated energy balances which show significant reductions of CO2 emissions. Results of RDR emission calculations are presented together with process energy savings and new heat integration in plant scale.

Pilot construction

RDR Pilot plant in Geleen will be built to test and develop RDR -technology process as a whole.

The program consists of 8-different type of test runs with about 50 start/stops divided over the 4 basic raw materials. Duration of a test is from few hours to two weeks naphtha run. Average hydrocarbon feed is 400 kg/h. In total, an amount of 450-ton hydrocarbons will be steam cracked. Project will function as a step towards commercial pilot size RDR eReactor. The project will deliver design tools for the commercial pilot phase, which will have industrially relevant size and will be connected to a running ethylene cracker. Air testing of RDR reactor power train is made in Finland during the construction of pilot and delivered for plug in to the Pilot plant.

RDR eReactor technology

Coolbrook RDR (RotoDynamicReactor - a turbomachine for cracking hydrocarbons) technology is an innovative technology to provide significant reduction of CO2 emissions by electrification of steam cracking. Coolbrook has developed concepts for the CO2 reduction and presents estimates for CO2 reductions of two concepts to replace conventional furnace section with RDR technology. In addition current status of the piloting project is presented. RDR reactor technology was first time presented in EPC 2018 conference

CO_2 reduction by RDR eReactor technology

The United Nations, European Union and, Biden administration and various multinational organizations around the globe have made battling climate change a focal point in their policy. Industry has also changed or is changing strategies to significantly reduce emissions of greenhouse gases (GHGs). These strategies are now converting to technology development projects. Emissions are hard to avoid as the materials and chemicals of petrochemical industry products are essential inputs to all major value chains. However, most stakeholders aim for the low-carbon transition to modernize and reinforce competitiveness on the world stage.

Incremental improvements on the current technology (e.g. furnace design improvement, process control and measuring device development) will not be enough to meet the 2050 targets. In order to be successful, new cracking technology is ideally electrified, with potentially no CO_2 emissions, flexible so that renewable energy production fluctuations can be handled and can be implemented at existing crackers sites into the existing infrastructure. In this way both the environmental and commercial viability would be maximized.

RDR technology is a step change and can be compared to start of Tesla saga. Comparing RDR technology to conventional steam cracking furnace is like placing a Tesla car and a car from 1990's on the same line.

Coolbrook has evaluated process concepts for furnace section and calculated plant-wide energy balances for both conventional steam cracker plant and steam cracking plant using RDR eReactor. Comparison of energy efficiency in RDR and conventional cracker is not straightforward. In conventional cracker, energy balance is typically calculated as fuel energy (primary energy) minimizing expensive electricity consumption. In electrification of steam cracking, the electricity consumption is to be dominant form of energy. Renewable electricity is a preferred source, but depending on location, will possibly be limited in supply in the near future. Another important aspect is that RDR technology does not change chemistry. Steam cracking is and will continue to be a significant energy consumer due to endothermic reactions and the high temperature requirement.

Considering these restrictions, plant-scale energy balance comparisons of conventional steam cracker and RDR technology have been made by Neste Engineering Solutions. Conventional steam cracker and RDR eReactor simulations were made using naphtha feedstocks and yields – assuming the separation schema and battery limits to be identical in all cases. Balances for several heat integration concepts using RDR eReactor have been evaluated. Losses are not included in balance. Comparison of two different steam cracker plant configurations using RDR eReactor against conventional steam cracker plant with similar feed production capacity are presented.

Plant-scale energy and material balances were simulated using Aspen+ Simulation package. Battery limits and conditions for energy balance are shown in Figure 1.

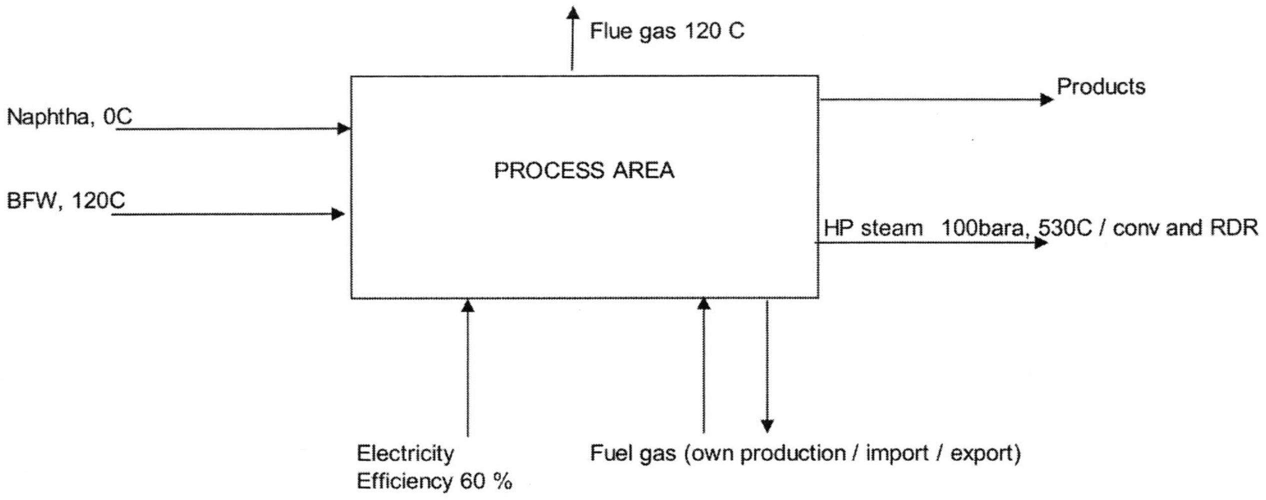

Figure 1. Battery limit for energy balance.

For both units following basis were used:

- Same process configuration, driver, pump and compressor efficiencies and product specifications are generally used for both units to allow comparison. The separation section configuration was not optimized regarding energy or investment cost.
- Conventional Frontend Demethanizer unit was modeled to yield material and energy balance of the separation section.
- For material balance purposes it was assumed that 50 % of MAPD (Methyl acetylene and Propadiene) is hydrogenated to propylene and 50% to propane.
- For material balance purposes it was assumed that 50% of acetylene yields ethylene and 50% is hydrogenated to ethane. Acetylene converters are located between high pressure Deethanizer and C2-splitter.
- Cooling water was made in cooling tower and the circulating water temperature was fixed at 24°C with 10°C temperature increase. Preliminary cooling water balance was made from ISBL systems.
- Diene hydrogenation of pygas (DPG-1) was not included.
- Quench water loop was used for C3-splitter reboiling and other process duties as much as possible
- Temperature of furnace effluent after TLE/steam generation was 370°C
- HP pressure steam was 100 bar and 530°C
- Inlet/cross ower temperature to radiant section/eReactor was 650°C
- Only superheated HP steam was taken as energy credit in energy balance

- Steam balances were based on same concept. Extraction temperature or back pressure for RDR was optimized. Steam turbines were replaced by electric motor in RDR technology cracker as needed due to smaller superheated HPS production

- Balances were calculated for 8400 h /year

- Steam to Hydrocarbon ratio was 0.5

Energy balance was calculated as follows:

- Energy consumed for heating and generation of electricity from fossil fuels (fuel gas) was calculated as fuel gas energy. Electric generation efficiency from fossil fuels 60 % (high efficiency gas turbine)

- Renewable electricity was calculated as consumed
- Fuel gas export/import was not taken as credit

- Three cases for electric consumption were calculated in balances:
    - no renewable electricity
    - 50 % of consumed electricity was renewable
    - 100 % of consumed electricity was renewable

Material balance was calculated for an ethylene plant producing 1 million Mton (metric ton) ethylene/year. Material balances are shown in Table 1. Typically, all kinetic models have been validated for conventional steam cracker so there are uncertainties in used kinetic model. RDR eReactor has different flow profile and high temperature, extreme mixing at the exit of the diffuser blades. Therefore, experimental data for RDR eReactor models (and kinetics) are required. However, it is well known fact that high temperature and fast heating rate/short residence time favors high olefin yields. Due to RDR eReactor's unique heating principle to convert kinetic energy of the gas directly to heat, it is possible to control heating rate from extreme short time to longer time as needed for optimum yields. RDR eReactor's high heating rate is not possible in conventional steam cracker furnace as heating take place through coil tube walls.

Once through yield for conventional steam cracking was taken from literature (Ullman Encyclopedia). In Ullman, the initial boiling point for conventional steam cracker naphtha feed was 26°C and final end point was 183 °C. In case of RDR eReactor naphtha feed, estimated initial boiling point was 32 °C and final end point was 158°C. Coolbrook recognizes small difference in naphtha quality and note that the literature yields are conservative compared to modern steam crackers. On the other hand, higher severity increases energy consumption while higher yields decrease it as per ton of products. For RDR eReactor yields were simulated by using kinetics developed by Neste Engineering Solutions. In kinetic simulations maximum operating outlet temperatures in RDR eReactor were 962°C and 923°C and residence times 10 and 35 ms. Operating pressure was 2 bar. For RDR eReactor ethylene yield was 38.3 % and propylene 13.6 % and for conventional steam cracker ethylene yield was 30.8 % and propylene 13.9 %. For recycle ethane and propane same conversion/yields were used for conventional and RDR eReactor. Olefin and hydrogen product qualities were same ((hydrogen 99.9 mol-%, ethylene 99.9 mol-% and propylene 99.5 mol-%). Using these once through yields RDR eReactor naphtha consumption is about 16 % less than conventional steam cracker for the same amount of ethylene produced.

	Conventional steam cracker	RDR eReactor
	kt/a	kt/a
Hydrogen	20	17
Fuel gas	504	341
Ethylene	1000	1000
Propylene	415	342
RawC4	234	176
Pyrolysis gasoline	560	471
Pyrolysis fueloil	118	23
Feed naphtha	2851	2370
Recycle Ethane	209	137
Recycle Propane	27	32
Total incl recycles	3087	2538

Table 1. Plant material balance comparison

Revamp/replacement of furnace Concept 1: Radiant and convection sections replaced by RDR eReactors

In concept 1 the radiant section of cracking furnace is replaced by RDR eReactor (Figure 2). Concept uses a preheater furnace for superheating dilution steam/naphtha to 650°C before it is fed to RDR eReactor. Heat integration is arranged so that primary quench exchanger, TLE1 generates high pressure steam which is superheated in secondary quench exchanger, TLE2. Fraction of the generated high-pressure steam is used for superheating dilution steam/naphtha mixture before it is returned to preheater furnace. Summary of results are presented in Table 2.

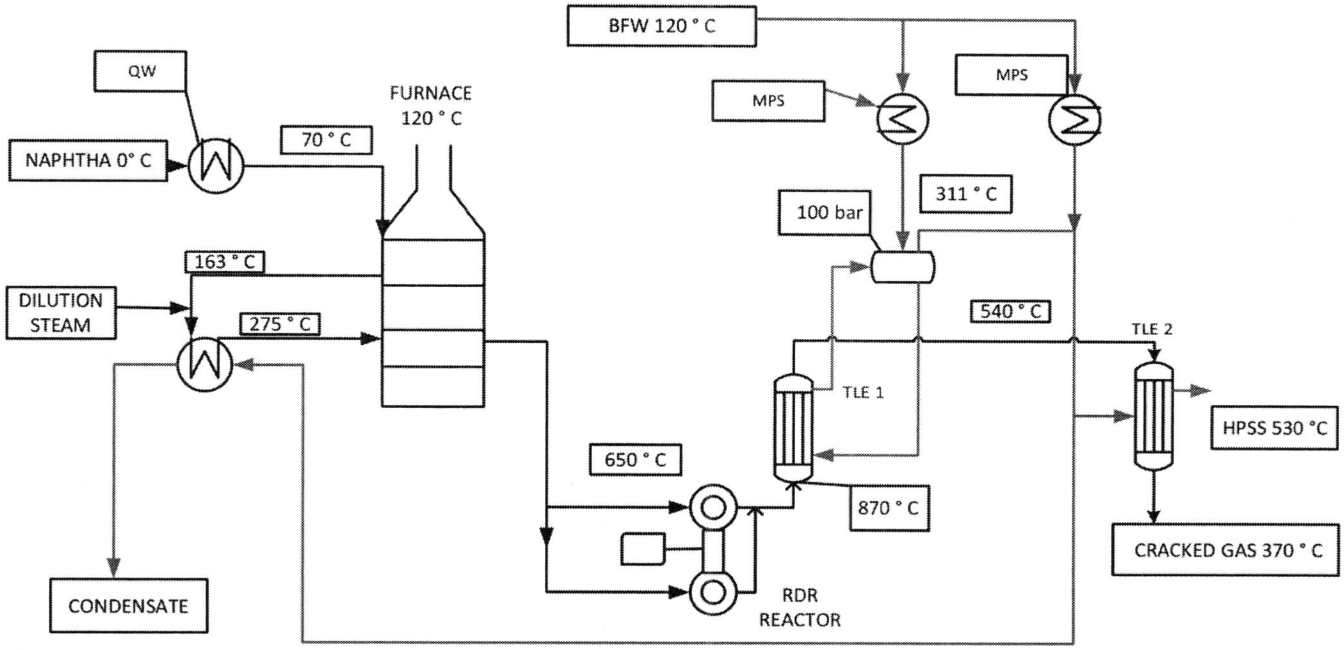

Figure 2. Concept 1 with Radiant section replaced by RDR eReactor.

Summary of the results are:

- Concept is easy to implement - in existing steam cracker plant e.g. by revamping one of the old furnaces for preheating and have several parallel RDR eReactors.
- CO_2 emissions compared to conventional steam cracker are 5 % less when 0 % of electricity is renewable.
- Significant reduction in CO_2 emissions is achieved when availability of renewable electricity increases.
- Preheater CO_2 emissions are about 30 % of conventional furnace emissions.
- Replacement gives significant size and complexity reduction of furnace section, releasing plot space for future expansion.
- Cooling water duty requirement (and heat loss to atmosphere) is about 25 % less than in conventional cracker
- NOX emissions decrease together with CO_2 emissions
- Boiler Feed Water consumption is about 50 % less in RDR eReactor (el. production excluded). However, also HP STM production is reduced

Primary energy summary	Conv	RDR	RDR	RDR
Amount of renewable electricity of consumed electricity	0	0	50 %	100 %
Consumption	MW	MW	MW	MW
Fuel fired in preheating/cracker furnace	776	218	218	218
Fuel needed for electric generation	12	547	273	0
Renewable electricity	0	0	164	328
Total energy consumption (fuel and renewable electricity)	788	765	655	546
Credit				
HP steam export (100bar)	100,5	4,1	4,1	4,1
Net energy consumption	688	761	651	542
Specific net energy consumption (inc. renewable electricity)				
GJ/t Ethylene	20,8	23,0	19,7	16,4
Utilities and CO2				
CO2 ton/ton ethylene	1,24	1,17	0,75	0,33
Cooling water duty, MW	419	312	312	312
Boiler water , t/h	484	242	242	242

Table 2. Concept 1 with Radiant section replaced by RDR eReactor. Result summary

CO_2 free cracking Concept 2: Fully electrified steam cracker

In Concept 2 (Figure 3) ethylene plant does not have fuel fired preheater furnace. Naphtha is preheated and evaporated using quench water/quench oil and medium pressure steam. High pressure steam is produced in TLE1. Part of generated high-pressure steam is used to superheat dilution steam/naphtha mixture before it goes to secondary TLE2 and for final heating to electric preheater. Saturated high-pressure steam is consumed in separation section. In the concept calculation medium pressure steam was imported for heating process streams. Steam turbine drives in compressors in separation section are replaced by electric motors.

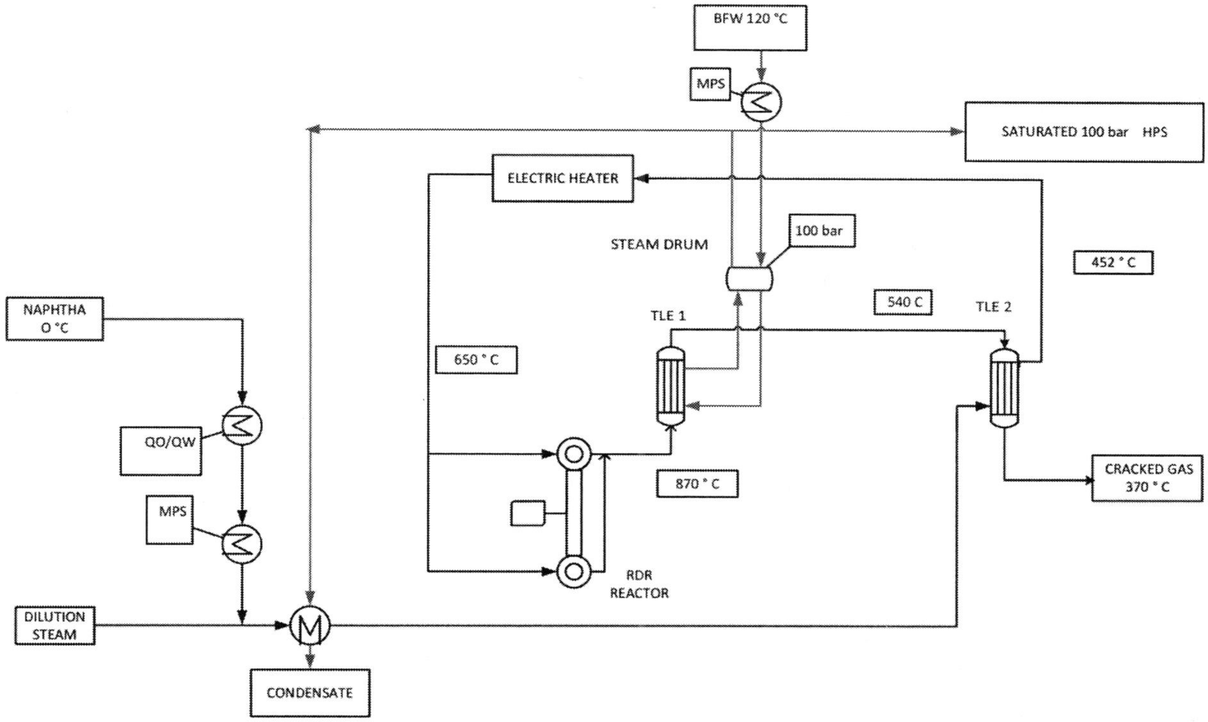

Figure 3. Concept 2 with Fully electrified steam cracker.

Summary of the results for a fully electrified steam cracker plant with RDR eReactors is shown in Table 3 below.

Primary energy summary	**Conv**	**RDR**	**RDR**	**RDR**
Amount of renewable electricity	0	0	50 %	100 %
Consumption	MW	MW	MW	MW
Fuel fired in preheating/cracker furnace	776	0,0	0,0	0,0
Fuel needed for electric generation	12	742	371	0,0
Renewable electricity	0	0	223	445
Fuel for MP steam production		47	47	47
Total energy consumption (fuel and renewable electricity)	788	789	641	493
Credit				
HP steam export (100bar)	100,5	0,0	0,0	0,0
Net energy consumption (including renewable electricity)	688	789	641	493
Specific net energy consumption (inc. renewable electricity)				
GJ/t Ethylene	20,79	23,87	19,38	14,89
Utilities and CO2				
CO2, kg/h (Total Fuel fired for including electric production and MP)	1,22	1,21	0,64	0,07
Cooling water duty, MW	419	262	262	262
Boiler water , t/h	484	251	251	251

Table 3. Concept 2 with Fully electrified steam cracker. Results summary

Summary of the results are:

- RDR eReactor concept is compact and easy to implement to existing steam crackers for capacity expansions or replacing existing furnaces.
- Electric preheaters can be parallel to preheater furnace to balance variation in renewable electric supply.
- No CO_2 emission from inside the battery limits
- CO_2 emissions are same when electricity production is included (60 % efficiency)
- Cooling water duty 38 % less than in Conventional Furnace (el. production excluded)
- Boiler Feed Water consumption 49 % less than in conventional
- Heat load to environment also 38 % less than in conventional
- Simple low CAPEX concept

Conclusions for RDR eReactor concepts

Calculation examples of these concepts clearly indicate the potential of RDR eReactor concept for electrification of steam crackers and reduction of CO_2 and NO_x emissions. RDR eReactor offers flexible concepts to optimize hot section heat integration. High temperature heat is used efficiently in hot section and less high-pressure steam is used in separation section or exported. Back pressure steam turbines and electric motors can be used in separation section instead of less efficient condensing turbines. Heat integration of feed hydrocarbon, dilution steam and boiler feed water can be optimized. This allows optimal use of high-, medium- and low-pressure steam, condensate, quench oil, quench water and "waste heat" temperature sources

Decreasing coke formation in steam cracker furnace coils has been one of the main areas of development for improving conventional steam cracker. One of the targets has been to decrease the surface temperature by the shape of coil or with different baffles in process side of tubes. In RDR eReactor fluid itself has the highest temperature which is less than the maximum surface temperature of conventional coil. Although aerodynamics is well understood and can be modelled with good accuracy and comprehensive kinetic models are available there are still uncertainties and risks. Pilot test runs will show performance of RDR eReactor: actual yields, safety, reliability, severity of coking, method of decoking and needed mitigation methods for coke formation (reactor surface coatings etc.).

Fuel gas was not calculated as credit in examples. In fact, it is important form of energy and it could also be calculated as credit. According to material balance fuel gas heating value for RDR eReactor concepts is about 5000 GWh for 1 million ethylene Mton production (590 MW). Fuel gas (Tail Gas) is Sulphur free and has high heating value. Fuel gas can be compressed and purified (remove excess hydrogen) so that methane can be fed to NG network. This fuel gas would be a significant "domestic" methane source and decrease import of NG.

Figure 4. The size of RDR eReactor on the right compared to the radiant section of the conventional steam cracker on the left having the same capacity.

The installation of a RDR eReactor is relatively simple due to its modular structure and small size compared to conventional crackers (Figure 4). In RDR eReactor, preheating furnace is significantly smaller and less complex than Convection Sections required in a conventional furnace. By-product recovery processes, equipments and utilities can also be simplified due to the more favorable composition of the olefins produced in the RDR eReactor. The small size of the drive train components and reactor allowes to construct and do maintenance off-site and simply install in a “plug and play” manner. In the current design philosophy, motor size of RDR eReactor would be between 20 to 40 MW. This would mean that a one million Mton ethylene plant would have 7 to 16 RDR eReactors (without spare). This can help bringing production and maintenance cost down as well as speeding up the installation process. As an example, in Figure 4 radiant section of conventional steam cracker is compared to size of RDR eReactor. Both units have the same capacity.

Status of the Pilot

Coolbrook will build and run a Pilot for the RDR technology at the Brightlands Chemelot Campus in Geleen, Netherlands for:

- Testing steam cracking in RDR eReactor with naphtha and LPG feedstocks,
- Validate and further develop reactor models,
- Test and improve reactor design and
- Demonstrate feasibility and safety of the RDR eReactor.

Coolbrook has earlier built and run a small RDR bench-scale reactor for the proof of concept. Couraged by the results of this proof of concept phase, now the 400 kg/h hydrocarbon RDR Pilot plant will be built in Geleen to test and develop RDR eReactor -technology process. In total, 450-ton of hydrocarbons will be steam cracked. Project will serve as a step towards commercial pilot size RDR eReactor. Coolbrook has chosen Brightlands because its excellent facilities, available technical resources, and location at the heart of European ethylene production.

Main parts of the Pilot are (Figure 5):

- Powertrain (motor, inverter, gear box, lubrication system and RDR eReactor)
- Transfer Line Exchanger (TLE) complete with STM Drum
- Flare
- Analyzers
- Feed station
- Preheating system
- Control system
- Pilot plant will be built to meet normal industry safety and emission standards

Preliminary design of the Pilot has been completed (FEL-study) and contractor will be selected by spring of 2021. Building the pilot will start in autumn 2021. The RDR Powertrain will be built in Finland for air testing. The air tested Powertrain will be transported to Geleen and connected to pilot plant early 2022. Test runs with hydrocarbon feedstocks starts in summer 2022 and will be comleted in autumn 2023.

Main technology risks are that RDR is not capable to achieve high yields, coking is more severe than expected and reliability, availability and operability are not on the same level or better than in conventional furnace. These issues have been noted in the design of Pilot by extensive CFD modelling of aerodynamics and using applicable standards in mechanical design.

The Pilot test program will be carried out in two phases and several sub-phases. The first phase will focus on starting the cracking process, maintenance, overall reliability and safety. After confirming safe operation, tests will continue to collect data for kinetic and aerodynamic modelling with naphtha feedstock.

The second phase will focus on collecting data modelling and testing product yields with liquefied petroleum gases (LPG) as a feedstock, overall reliability, and safety.

Results will be analyzed and used to improve the RDR technology with the help of leading scientists in the fields of aerodynamics, fluid dynamics and chemical modelling from Cambridge, Oxford, and Ghent universities.

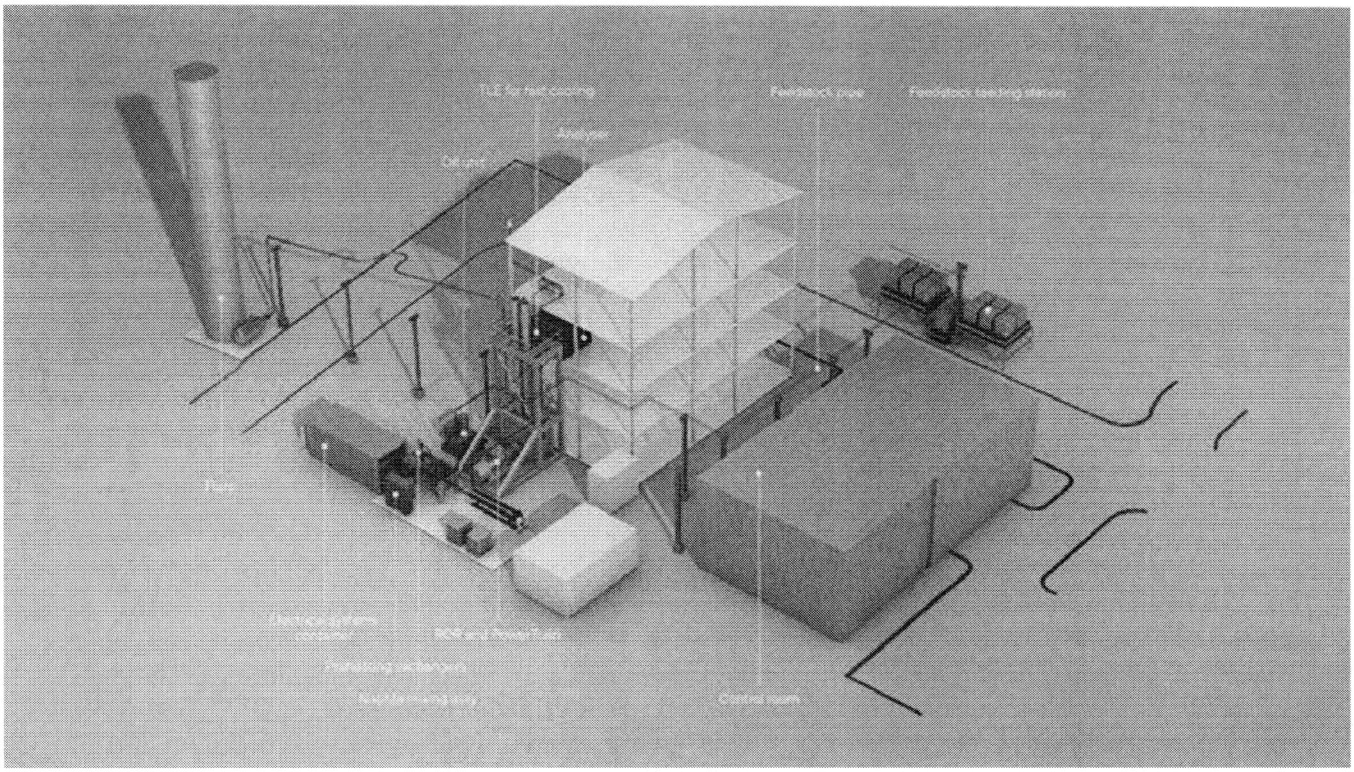

Figure 5. RDR eReactor pilot in Brightlands campus.

616879 Management Systems and Culture Important for Sustained Operations

Abstract:

Currently, the Department of Energy (DOE) stores approximately 56 million gallons of radioactive and hazardous waste in 177 underground tanks at the Hanford Site. The waste was generated from production of plutonium for the United States' nuclear weapons program starting in the second World War and finishing in 1990. Approximately 20 million gallons of that waste is in a liquid form (supernate), approximately 10 million gallons is in the form of insoluble sludge materials, and the remainder is in a partially soluble solid form referred to as saltcake. Treatment of the tank waste is planned with the Waste Treatment and Immobilization Plant (WTP) being the principle plant where this will be accomplished. The WTP is currently being constructed and large portions of the plant are already starting up. Waste treatment commissioning and operations are planned to start in the early 2020s.

The start of waste treatment operations will mark a significant event in Hanford history. Not since the 1980s, or at least a generation of engineers, will the Site experience the sustained operations of a multi-functional chemical process plant. This situation makes for an interesting study in transitioning management systems and culture from one supportive of design and construction to one conducive to sustained operations. Making ready for safe operations and then to demonstrably show readiness in Operational Readiness reviews and assessments is a well-defined process in the DOE Environmental Management complex. However, there is little programmatic guidance on how to make ready and demonstrably show readiness for sustained operations.

The development of a Sustained Operations Readiness Program was conducted based on the author's professional experience, input from specific individuals with relevant expertise and the author's review of public information. A Sustained Operations Readiness Program is not intended to replace or in any way diminish the role for Operational Readiness Reviews and Assessments as required by the Department of Energy. Rather, the intent is to build upon the Operational Readiness concept to maximize readiness for sustained operations through commissioning and into operations, and to establish a culture of simultaneous operations excellence in production, safety and environmental management. The work has concluded that the state of sustained operations for a complex processing plant can be objectively characterized by 12 features. These features represent the conducive culture or conditions in which the plant, paper and people interact to produce a sustained operations state. The 12 features were further consolidated into the five programmatic and executable elements described below with associated recommendations specific for Hanford waste treatment operations:

Risk management
Risks and opportunities should prioritize continuous improvement initiatives and be informed by the design and technical authorities and stakeholder monitors.
For Hanford waste treatment operations, an integrated risk management program led by the DOE is recommended.

Continuous improvement
All levels of the project organization should exhibit commitment to continuous improvement and learning.
At Hanford, the tank farm contractor had introduced the Lean management system as a continuous

improvement program. The program has yielded efficiencies in tank farm operations and waste treatment operations planning and introduced a culture of continuous improvement. A waste treatment operations program-wide continuous improvement system is recommended for Hanford waste treatment operations.

Continuity of design and technical authorities
The authorities are responsible for managing knowledge for its effective use in the other elements. Recommended for Hanford waste treatment operations is a body of technical program managers from the key technology providers chartered to provide the technical authority and a continuing role for the WTP Engineering, Procurement and Construction contractor as the design authority.

Stakeholder monitoring
Active and deliberate monitoring of stakeholder relationships.
Hanford waste treatment operations is characterized by complex and wide-ranging stakeholders at Federal, State and local levels of government and regulation and various public bodies. Centralized and integrated stakeholder monitoring and management is recommended as an organizational element.

Sustained operations integrated business planning
Business planning integrates and balances technical, financial and regulatory objectives. In a commercial venture this is arguably a business plan.
An integrated Hanford waste treatment operations business plan is recommended for the Site to most effectively compete with other sites in the current Federal environment competitive for Congressional funding.

Readiness evaluation for sustained operations is a continuous function that progresses from overseeing establishment of the five elements during design and construction phases to monitoring the output of those elements performing during commissioning and operations. Continuously assessing readiness for sustained operations becomes an important business function for the Department of Energy's Office of River Protection as competition for Federal funds increases with increasing waste treatment operations throughout the Environmental Management complex.

621580 Carbon Leadership in a Low Carbon World

Abstract:

Carbon dioxide levels in the atmosphere have increased from about 275 ppm prior to the industrial revolution to over 415 ppm today. Nature cannot keep up, and there is now over 2 trillion tonnes of CO_2 that has accumulated in the atmosphere. This presentation will touch on point source emissions capture but will focus on Direct Air Capture technology coupled with subsurface sequestration (DAC+S). A technology overview and a summary of potential business models will be discussed.

620745 Public speaking for ChemEs: improve your oral communication skills

Abstract:

Whether you are giving a talk for a class, at a (virtual) conference, or to get a job, it is essential that you shine in every way. Learn the essentials of giving an excellent speech every time. Discover what a speech says about you, how to prepare for a speech, how to combat anxiety, and how to show confidence and poise. Speaking, articulation, and projection techniques, interaction with the audience, and specific presentation issues for scientists/engineers and for science-based conferences will be addressed. And of course, since 100% of our oral presentations are via virtual means, we will delve into the logistics and considerations we can keep in mind as we speak via webcam and present via zoom.

SPRING21 +17TH GCPS
A Joint AIChE and CCPS Meeting

"Hello? Can Everyone See My Screen Now?" – Conducting an Effective PHA from a Distance

Lauren Moyer, CCPSC
Principal Chemical Engineer
Global Process Safety
Eastman
P. O. Box 1972
Kingsport, TN 37662
lmoyer@eastman.com

Prepared for Presentation at
American Institute of Chemical Engineers
2021 Spring Meeting and 17th Global Congress on Process Safety
Virtual
April 18 - 23, 2021

"Hello? Can Everyone See My Screen Now?" – Conducting an Effective PHA from a Distance

Lauren Moyer, CCPSC
Principal Chemical Engineer
Global Process Safety
Eastman
P. O. Box 1972
Kingsport, TN 37662
lmoyer@eastman.com

Keywords: Hazard Identification and Risk Analysis, Process Safety, Safety & Health

Abstract

As the ability to travel and on-site access to chemical manufacturing facilities disappeared in early 2020 due to a global pandemic, operating companies have had to find new ways to complete Process Safety Management (PSM) activities including Process Hazard Analyses (PHAs), audits, and incident investigations. These activities typically rely on personal interactions with manufacturing staff and field inspections of the equipment and facility siting. Although crucial risk management activities could be completed with a "just get it done" attitude to weather the current work climate, such an approach could potentially result in gaps in risk assessment and mitigation. How can we do better? What best practices for virtual PSM activities, forced upon us due to external circumstances, could permanently increase our work flexibility, reduce travel costs, and improve our efficiency?

This paper will consider some of the potential pitfalls and challenges in conducting risk management activities virtually, focusing on Hazard Identification and Risk Analysis (HIRA) techniques such as PHAs. These include practical issues, like communications technology and conducting field walk-downs, and softer skills, like effective facilitation, managing team dynamics, and ensuring effective contributions of all participants. When PHAs require multiple lengthy sessions and a high level of interaction with all team members, maintaining team focus can be more challenging in a virtual environment. Effectively collecting and distributing the study's Process Safety Information (PSI) to all team members may require both creative methods and new technology. In-person field tours of the process may be almost impossible, but could a thoughtfully curated collection of photos or videos be an effective substitute? Tools and best practices developed at one operating company to make virtual PHAs less bad and less ugly, and maybe even better in the long-term as a result of these adaptations, will be presented and discussed.

1 Introduction

The regular assessment of a chemical process' hazards and risk is a cornerstone of Process Safety Management (PSM). The Process Hazard Analysis (PHA) has been recognized as essential to safe operation of chemical process sites by both industry organizations such as the Center for Chemical Process Safety (CCPS) and regulatory bodies around the world.

High quality process hazard analyses (PHAs) are built upon in-depth discussions of the technical details of a process by knowledgeable individuals to identify hazards and then determine if sufficient safeguards are in place to mitigate them. The chief role of a PHA study leader is to effectively lead these discussions and obtain team consensus on the process risk and any needed recommendations. When outside factors, such as a global pandemic, disrupt the ability to hold these discussions and other "normal" aspects of work, it is tempting to just do the minimum regulatory requirement in order to weather the storm of travel limitations and restricted access to chemical facilities. Is it possible to delay PHA studies? What is the minimum revalidation methodology that can be employed? What is the smallest PHA study team that is required?

This philosophy is problematic for several reasons. The most immediate concern is that hazards in a process go unidentified due to delaying a study, conducting a less thorough evaluation, or not having enough knowledgeable personnel available in the study. Any one of these situations could potentially lead to an increase in process safety incidents at the facility. An organization should want to optimize performance of PHAs, even during difficult situations, so gaps in risk assessment and risk management do not develop. A long-term concern is that this approach could erode the commitment to process safety excellence in an organization. If, when the going gets tough, risk management activities such as PHAs are compromised, it could send a message to the organization that these activities are not of long-term importance to management.

An old proverb states that, "Necessity is the mother of invention." [1] A global pandemic requiring physical separation of people could ultimately lead to permanent improvements in how work is performed, including conducting effective PHAs. The ability to conduct an effective PHA from a distance allows for a critical risk management activity to be completed in a safe manner in the near-term due to pandemic restrictions. In the future this ability could allow for a long-term reduction in money or time spent traveling to sites and an increase in work flexibility for employees. By evaluating facilitation techniques and technology to allow for remote work, an organization could find opportunities to improve both the tools and methods used to conduct any PHA study, not just the ones conducted at a distance.

2 The Challenges of Remote PHAs

Conducting PHAs with some, or all, of the team participating remotely has several challenges when compared with having everyone meet face-to-face. Team members might have limited access to the chemical facility and its equipment, people, and information. Being able to see the equipment and the operating environment is crucial to understanding potential hazard scenarios resulting from equipment placement, facility siting, or human factors issues. Having easy access to personnel and information during the study will result in a more thorough understanding of the process design and operation, with fewer recommendations to follow-up on questions from the team after the study.

Remote meetings are extremely reliant upon reliable and effective technology for success and can be limited by the shortcomings in technology. If all team members do not have access to necessary software, hardware, and network capabilities, it will quickly become difficult to keep the team together in the discussions. Access to requisite technology is not the only concern; if all team members do not have competency in using collaboration technology, the study meeting will become mired in the frustration of waiting on participants to get connected or find and display the correct resources.

PHA team meetings are mentally taxing, even under the best conditions. With the limited visibility of remote meetings, it can be difficult to "read the room" to detect and address fatigue, distraction, and other dysfunctional meeting behaviors. It can be very difficult to convey detailed chemistry or engineering information effectively and keep all team members engaged in the discussion for the duration of the study.

All of these issues can set up a perfect storm of inefficiency, miscommunication, and frustration which could derail the PHA, resulting in a less thorough risk evaluation for the process. Can guidance and tools be developed to address these issues before they disrupt the PHA study?

3 How Improvements are Made

With the established need for conducting high quality remote PHAs, companies and process safety professionals must rise to the occasion and develop methods, tools, and guidance to aid in the success of its PHA teams. Learning and development specialists generally recommend improvements be based on a mix of three areas: on-the-job learning, learning from others, and formal learning events [2]. This model can be applied to learning and developing PHA leadership methods as well. When faced with the challenge of shifting to remote work in early 2020, Eastman has used these techniques to develop guidance and tools for leading successful virtual PHAs.

3.1 Formal Learning Events

In the initial stages of the 2020 global pandemic, many process safety consulting organizations were offering training and services to assist with conducting risk management activities remotely, including PHAs. Even for operating companies who facilitate their own PHAs, learning from the experience of others in the industry through webinars, papers, and training classes is an excellent way to broaden horizons and increase knowledge. Participating in training from a variety of external resources can offer different viewpoints, and the trainee can then determine which ideas would be a good fit for application within their organization.

3.2 Learning from Others

Most large, global operating companies have many personnel performing PSM activities, often located in a variety of geographical locations and supporting facilities of varying sizes. Tapping into this variety of experience can help the entire organization learn and improve. With remote work becoming more of a certainty as the 2020 pandemic increased in scale, it became necessary to figure out a way to conduct PHAs with limited access to the facilities and required personnel. To jumpstart these efforts, a brainstorming session was held with experienced study leaders and those knowledgeable in process safety to assess the roadblocks of virtual PHAs and to recommend solutions for overcoming them. The results of this effort included a guidance document for remote PHA best practices and clarification of existing processes and procedures to reflect the remote work environment.

3.3 Learning from Experience

Seeking advice and learning from others is important, but often personal experience is the best teacher. With nearly a year of experience in leading remote PHAs in a variety of constraints, dictated by local pandemic conditions, study leaders in this global company have improved with practice and found ways to overcome barriers to success. In the interest of continual improvement, it is important to capture these learnings, pass them on to others, and determine if additional tools and guidance can be developed to further assist in PHAs conducted remotely, or even in more traditional meeting structures. Surveys directed at PHA study leaders and team members are a good tool to gather initial feedback, with more targeted follow-up occurring in discussion meetings. This follow-up can identify what is working well, what pain points still exist, and any new ideas for improvement not considered previously.

3.4 The Results of Intentional Learning

The combination of these three efforts resulted in improvements in three key areas of leading PHAs: study preplanning activities ("Be Prepared"), technology and logistics ("Be Flexible"), and meeting facilitation ("Be Intentional"). A guidance document discussing these key issues and possible solutions and best practices was developed and distributed to all PHA study leaders in the organization. Quarterly study leader networking meetings are an additional forum for continued discussion and information sharing on conducting PHAs virtually. These key improvement practices shown in Figure 1 are discussed in the following sections.

Figure 1: Keys to Executing Successful Virtual PHAs

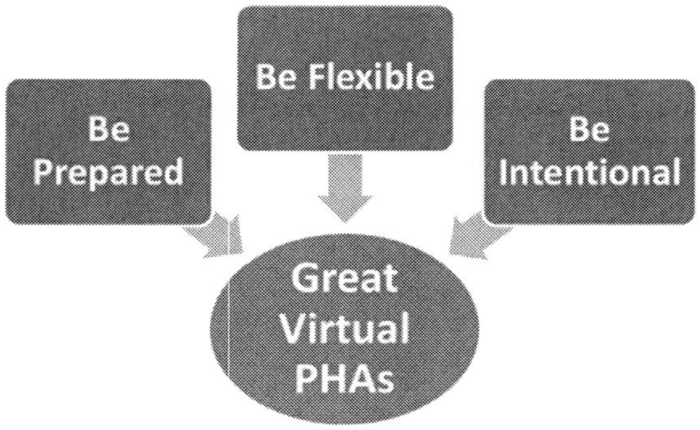

4 Be Prepared

Due to the detailed nature of PHA studies, pre-planning is a crucial aspect of any successful study. Information about the chemistry, equipment, and procedures must be gathered, reviewed by the study leader, and then distributed to other team members for reference. The boundaries and structure of the study must be determined or, in the case of revalidations, reviewed and reconfirmed. With remote work, access to and distribution of this information shifts to electronic formats. The study leader may require extra assistance in preparing for the study and need to plan for extra time to prepare for the added complexity a remote study brings.

4.1 Planning, Planning, Planning

In preparing for battle I have always found that plans are useless, but planning is indispensable.
- Dwight D. Eisenhower [3]

Planning is important for the success of any PHA. However, when the PHA is conducted remotely, additional issues must be considered in the preparation phase. The agenda and schedule will require additional thought due to the potential for participants being in different time zones and the increased mental fatigue associated with virtual meetings. The responsibilities and interactions of the study leader and the scribe during the meeting will need to be planned. The study leader and scribe should work together to decide who will be the virtual meeting "leader", determine what information should be shown on the shared screen during the study, and prepare additional documents or visual aids are needed for the study.

Ensuring all team members have access to the process information as well as the hardware and software tools needed to participate virtually is an issue unique to remote PHAs. This will require additional thought and preparation beyond routine PHA planning. Team members should all be able to reliably see and hear the information being discussed ideally with minimum distractions. It is also important for all team members to have a working knowledge of the virtual meeting technology to be used for the PHA, so it may be necessary to provide information on available technology training and help documents or do a test run of the meeting set-up ahead of time.

4.2 Local Facilitator

Consider naming a local on-site study facilitator to assist with gathering information and conducting any on-site study coordination for the study leader. This person can serve as the "boots on the ground" in preplanning activities and be of particularly high value to a study leader located in a different geographical location from the facility with limited travel capability. They can assist with gathering process information, distributing it to team members, arranging technology needs for team members, and any other tasks best suited to an individual with access to the facility. This person can also help to arrange space for operations team members to join the meeting with minimum distraction, away from the noise of the control room or shared operator spaces (e.g. a conference room or empty office). This individual could also conduct a virtual plant tour to gather photographs or videos of equipment and key operational tasks to share with members of the team located remotely.

4.3 Field Walkdown Checklist

One of the most significant issues identified in brainstorming efforts was the lack of access to the facility for a tour of the operating area being studied. The field walkdown can give crucial insight into how equipment is arranged, how operators interact with equipment when performing their tasks, facility siting concerns, and potential human factors issues in the process. The question then becomes, can a photographic or video tour of the facility also convey this important information? Photos with the context provided by operations representatives familiar with the area are a great way to supplement discussions in PHA sessions. To better direct the local facilitator conducting the tour, a checklist of items important to hazard evaluations was developed. See Table 1 below for an example tour checklist containing a list of key items.

Table 1: Example PHA Field Tour Checklist

Overall Condition of the Plant

- Major pieces of equipment
- Congestion, housekeeping, and general egress paths of routine task areas
- Ventilation issues
- Lighting issues
- Drainage and spill containment areas
- Proximity to other processes, boundaries, and occupied buildings
- Deteriorated equipment (concrete spalling, erosion, corrosion, etc.)
- Inadequate equipment support (sagging piping, noticeable vibration, etc.)
- Equipment and piping labeling

Safety Equipment

- Fixed fire protection systems (coverage, activation points) and fire extinguishers
- Safety showers and eye bath (SS/EB) locations
- Combustible or toxic gas analyzer locations
- Presence/condition of grounding cables
- Pressure relief devices and their discharge locations
- Proximity of flares, furnaces, or other ignition sources in the area

Operator Interface Locations

- Tripping or falling hazards
- Accessibility of key valves, filter housings, or other frequent manual tasks
- Piping blocks access to equipment
- Sample locations (egress, safety shielding, grounding, access to SS/EB, etc.)
- Awkward or cramped positions or other ergonomic concerns
- Unclear or confusing operator interface locations

Development of this field walkdown checklist is one example of an improvement which could apply to *all* future PHAs and not just those conducted virtually. Previously, this formal guidance did not exist, and the ideas of what to look for on a tour were based on study leader experience and judgement. This checklist can be used as a training tool for new study leaders, team leaders, and less experienced PHA teams. This is an example of how reexamining existing processes and procedures during short-term disruption or change can reveal gaps and lead to permanent improvements.

4.4 Increased Reliance on Software Tools

In addition to using computer tools for conducting the PHA meetings, other software tools also can help with planning for a virtual PHA. With limited access to office facilities in the 2020 pandemic, access to copiers and printers became more difficult. This resulted in a shift towards using software tools, such as drawing management software packages, to annotate drawings and procedures with the PHA nodes and other notes and concerns. Collaboration software made it possible to share the annotated drawings and procedures with all team members without relying completely on hardcopy versions of key process information. Tutorials on using these tools were offered to study leaders to help jump-start their use. This is another example of an improvement in work processes that could pay long-term benefits, with reduced paper and copier use and easier electronic archiving of PHA documents after study completion.

5 Be Flexible

Conducting meetings in the virtual environment requires flexibility. With a reliance on computer and network technology, meeting logistics can require contingency planning to anticipate unexpected issues. Consider having a back-up plan for network availability, such as a wireless hotspot, or have a secondary location available for team members to use if the computer or network is not cooperating or if excess noise or distractions are an issue. Also consider having a list of information technology (IT) resources and contacts available in anticipation of any problems.

The schedule of the PHA should have enough flexibility in it to allow for potential downtime due to network or other technology issues. The team should be prepared to change meeting arrangements if things are not working well for everyone on the team.

The study leader should also be aware of non-technology issues, such as meeting fatigue, and be prepared to adjust the schedule to take more frequent breaks or shorten the meeting sessions if needed. Generally, limiting sessions between breaks to around 1.5 hours is good practice for virtual meetings, which may be slower than the generally accepted pace of an in-person meeting. Additionally, it may be necessary to adjust how the meeting is being conducted to ensure everyone is able to follow the discussion and that distractions are minimized. This may mean changing what is shared in the virtual meeting screen, such as using video versus projecting the PHA software or a drawing or procedure. Seeking frequent feedback from the team on what is working well and what adjustments are needed as the study progresses is also a good practice.

6 Be Intentional

One of the most challenging aspects of conducting PHAs remotely is managing team dynamics and maintaining good team engagement. PHAs are frequently multi-day or even multi-week reviews requiring a high degree of focus and deliberation, and keeping a team fully engaged is a challenge even in the best work environment. As noted above, planning and flexibility can help head off some engagement issues before they start. In the moment of the meeting, leading a PHA virtually requires a very intentional approach to facilitation to manage issues that emerge as the study progresses.

6.1 Calling on Everyone

When conducting an in-person PHA, it is often possible to "read the room" to determine if people might have something to add to the conversation or if more discussion is needed on a topic. Even with video capability, it can be very difficult to assess this in a virtual meeting. This requires the facilitator to be very intentional about seeking input from team members by asking specific team members questions about the equipment or the scenario being discussed. Frequently calling on people by name is also a good way to confirm people are still connected and have not encountered a technology issue.

6.2 How a Question is Asked Matters

When facilitating a virtual meeting, how a question is phrased or framed can greatly increase or decrease the level of interaction and discussion that follows. This is also true with in-person interactions, but it is even more crucial with virtual PHAs where the participants may not be able to see each other and respond to the energy in the room. In *The Secrets of Facilitation*, trying to word questions carefully to avoid silence is given as an important strategy for virtual meetings. "Avoid asking a question to the group that no one can answer." [4] Try to keep the questions and conversation within the scope of knowledge and influence of the PHA Team. Targeted, but open-ended discussion questions can get the group engaged faster than short yes or no questions, especially when combined with intentionally calling on people for input as discussed above. As a study leader, asking team members to talk through how a particular piece of equipment works, how a task is performed, or other details can get them involved in the discussion of what could go wrong in that part of the process.

6.3 Dealing with Dysfunction

When leading virtual PHAs, the combined responsibilities of managing the flow of the study and the meeting technology can mean potentially dysfunctional team behaviors go undetected more easily than in non-virtual meetings. In *The Secrets of Facilitation*, a number of negative meeting behaviors that can disrupt a meeting and its outcomes are listed and discussed. Table 2 shows the most common dysfunctional behaviors in meetings. [4]

Table 2: Common Meeting Dysfunctions [4]

Cell phone junkie / workaholic	Person is distracted by phone / computer work
Door slammer	Person leaves the "room" in disgust
Dropout	Person does not participate in the discussion
Interrupter	Person interrupts others or finishes their sentences
Late arriver / early leaver	Person habitually arrives late / leaves early
Loudmouth	Person dominates the discussion
Naysayer	Person is noticeably negative without offering solutions
Storyteller	Person likes to tell long-winded stories
Topic jumper	Person frequently takes the group off topic
Verbal attacker	Person makes negative comment about / directed at someone
Whisperer	Person holds side conversations during meeting
Low Energy (Group)	Energy in the "room" is low, participation is waning

Several of these items, such as workaholic, drop-out and low energy, can be addressed with the intentional interactions listed in the sections above. Others may be harder to either detect or discreetly address in the virtual environment. Being intentionally aware of the interactions occurring as the study progresses and their tone can help with this. Additionally, a local facilitator can serve as the "eyes in the room", especially if only the PHA leader is located remotely. They may also be able to brief the study leader on potential dysfunction tendencies of team members ahead of time. Establishing some key ground rules, such as "One Conversation at a Time" or "Respect the Speaker", for meeting conduct ahead of time will set expectations for the team. Considering possible ways to head off some of the more common issues early can help ensure a successful study.

While virtual meetings, including PHAs, can introduce additional challenges for the leader, the suggestions above show that a heightened awareness of facilitation techniques and intentional use of troubleshooting methods can produce a well-executed, collaborative evaluation of process hazards for the facility – even virtually.

7 Conclusions

While the global pandemic of 2020 is not an event anyone would like to see repeated anytime soon, in many ways it has forced a detailed, critical evaluation of what work methods are possible and what technology is available to connect people in new and different ways. Conducting PSM activities, such as PHAs, is no exception.

With intentional efforts to determine and address potential roadblocks to remotely conducted PHAs, the tools and procedures can be established to enable study leaders to prepare for and successfully lead studies in the virtual environment.

The keys to effectively leading a virtual PHA can be summarized with three ideas:

1) Be Prepared – diligent and thorough preplanning

2) Be Flexible – meeting technology and logistics contingency planning

3) Be Intentional – deliberate and thoughtful facilitation techniques

Implementation of these key practices can transform virtual PHAs from a necessary evil to an additional tool in the PSM toolkit to help achieve the ultimate result of risk reduction and improved performance at chemical facilities.

8 References

[1] "Necessity is the mother of invention." Merriam-Webster.com Dictionary, Merriam Webster. Available at https://www.merriam-webster.com/dictionary/necessity%20is%20the%20mother%20of%20invention. Accessed on February 2, 2021

[2] Training Industry. "The 70-20-10 Model for Learning and Development." Available at https://trainingindustry.com/wiki/content-development/the-702010-model-for-learning-and-development/. Accessed on February 12, 2021.

[3] Eisenhower, Dwight D. Available at https://www.azquotes.com/quote/87601. Accessed on February 19, 2021.

[4] Wilkinson, Michael. The Secrets of Facilitation. p. 189-196, 298. Jossey-Bass. San Francisco, CA. 2012.

SPRING21
+17TH GCPS
A Joint AIChE and CCPS Meeting

Exercising Operational Agility in Virtual Training Sessions

Stuart Chia
Principal, Sherpa Consulting Pty Ltd
Level 8, North Tower
1-5 Railway St, Chatswood, NSW, 2067
stuart.chia@sherpaconsulting.com.au

Simone Ligori, Angelyn Wu
Safety and Risk Engineers, Sherpa Consulting Pty Ltd
Level 8, North Tower
1-5 Railway St, Chatswood, 2067
simone.ligori@sherpaconsulting.com.au, angelyn.wu@sherpaconsulting.com.au

Prepared for Presentation at
American Institute of Chemical Engineers
2021 Spring Meeting and 17th Global Congress on Process Safety
Virtual
April 18 - 23, 2021

Exercising Operational Agility in Virtual Training Sessions

Stuart Chia
Level 8, North Tower
1-5 Railway St, Chatswood, NSW, 2067
stuart.chia@sherpaconsulting.com.au

Simone Ligori, Angelyn Wu
Safety and Risk Engineers, Sherpa Consulting Pty Ltd

Keywords: virtual, training, facilitator, learner

Abstract

As businesses, including the process industries, adapt to the 'new normal' in the re-entry and recovery phase of COVID-19, there has been a major reset in the way work activities are prepared, conducted and delivered. In this challenging environment, from a business continuity viewpoint, it may not be prudent or possible to have face-to-face workshops involving key personnel. This certainly applies in the conduct of hazard management training where personnel would typically learn, apply and reinforce required learning competencies in a collegiate environment.

It has become evident facilitators have had to alter the way they organise and deliver content in a virtual environment to ensure there is no adverse change in quality. The challenge is to maintain engagement and foster a nurturing environment to promote learning and communication amongst the learners as well as between the learners and facilitator(s).

This paper relays the good and bad experiences from Sherpa's delivery of hazard management virtual training sessions to operating facilities located nationally and overseas. From this experience Sherpa has identified 8 key steps in the plan and prepare phase of developing a training session that can be used to equip facilitators with operational agility during training sessions. By focusing on learner and facilitator profiling, preparing training material to suit a virtual environment and structuring sessions to maintain learner engagement, Sherpa has been able to develop solutions to overcome some of the challenges faced in a virtual environment. Together these lessons learned can ensure key required learning outcomes and learning experience are successfully delivered.

1 Introduction

1.1 How COVID-19 has changed the way we work

Almost a year has passed since the World Health Organization (WHO) declared COVID-19 a pandemic, with lockdowns forcing businesses to transition out of the physical workplace and adopt virtual communication platforms while working remotely.

Post COVID-19 some things may return to how they were, but many learnt behaviours from the pandemic will form the "new normal" with permanent changes to the way we live and work. There has been a major reset in the way work activities, particularly training sessions, are prepared, conducted, and delivered as businesses try to maintain business continuity in a volatile environment. In these unprecedented times, COVID-19 has accelerated work practices and highlighted the importance of technology, adaptability, and agility in the workplace. Namely, the biggest impacts have been:

1. Greater support for flexible working arrangements
2. Restructuring and rethinking of products and services
3. Integration of technology in the workplace
4. Greater focus on workforce effectiveness, communication, and trust

Adapting to this 'new normal' has meant a re-think in delivery of services to process and critical infrastructure industries. This certainly applies to training. Process safety training has always been essential for personnel involved in the design, operation, and maintenance of a facility, and perhaps even more so in the midst of a pandemic. It is important personnel are trained to be aware of hazards (what can go wrong?), understand existing controls (how can we stop a major accident from happening?), and can monitor they are effective (do we have the information and assurance to know controls are working effectively?). The risk of major accidents at hazardous facilities are ever present. And so, process safety training for personnel must also continue as one element of effective hazard management.

1.2 An opportunity for innovation

Conducting business during a pandemic is a challenge that brings an opportunity for innovation-led growth created by the changing landscape. Businesses have had to rapidly learn to be more agile, effectively manage internal communications and discover optimized ways to meet targets. A key business activity that has seen a shift as a result of these changes is training. In a world where face-to-face training has been minimized/halted, facilitators and learners have had to adjust to a 'new normal' of virtual training sessions. As facilitators, Sherpa recognizes that while virtual training provides businesses with increased flexibility and potential time savings, it will never replace the quality, interactivity, and engagement traditional face-to-face training offers. There is a need to bridge the gap between virtual and physical classrooms to enhance distance learning experiences necessitated by COVID-19. This presents an opportunity to identify new innovative solutions to adapt in-person material to facilitate safe, valuable, and optimized online learning within the limitations of current commercially available communication platforms. By identifying and understanding the key challenges associated with virtual learning, facilitators will be able to better address them and bridge the gap between virtual and face-to-face learning.

Sherpa's experience with virtual training sessions revealed two 'bad and ugly' challenges faced in this environment: (1) maintaining motivation and learner engagement in a virtual environment and (2) dealing with technology – in terms of functionality and useability. This paper further explores these challenges and delves into how Sherpa bridged the gap between face-to-face and virtual training through a strong emphasis on planning and delivery.

2 The ('bad and ugly') challenges of a virtual training environment

Traditional face-to-face training and virtual training share many similarities, learners are still required to achieve the planned learning outcomes of the session, complete exercises, and participate in group activities. Despite sharing these similarities, virtual training does not have the benefits face-to-face training sessions have, leading to new challenges for facilitators. Two differences which pose a challenge to facilitators are the need to communicate through a screen, and the use of new technology requiring computer literacy. This section will examine these two aspects and explore how they challenge facilitators and in turn require them to develop innovative solutions to deliver virtual training.

2.1 Communicating through a screen

A key difference between teaching virtually and in-person is the separation and lack of natural human interactivity between the learner and the facilitator. Communicating through a screen leads to a loss of connection and perceived lack of accessibility to the facilitator and other learners. As a result, it becomes difficult for facilitators to read non-verbal cues, if any at all, create an open learning environment and maintain learner engagement. Training is critical for all business operations; it is not uncommon for companies to spend significant sums annually on internal or external training. Although virtual training sessions have allowed for greater flexibility and accessibility, COVID-19 has pressured businesses to be aware of financial constraints and it becomes even more critical to ensure optimal training is delivered.

2.1.1 Non-verbal feedback

In face-to-face training, the facilitator can gauge the engagement of the class through non-verbal cues and immediately adapt their content, delivery and pace to better suit the class. Virtual learning however, naturally limits immediate feedback and non-verbal communication between the facilitator and the learner. This makes facilitating even more challenging. While some learners may participate more than others, the facilitator is unable to tell if the lack of participation is due to their introversion, high focus listening to the training, or not being engaged properly. The lack of feedback to the facilitator limits their ability to respond and decide on how to best pivot their content or delivery to better suit different learner mechanisms – such as if the pace may be too fast or if the level of vocabulary used may be too advanced for an English as Second Language (ESL) audience. The absence of feedback to the facilitator results in a lack of response from the facilitator and then generates further silence and possible confusion from the learners.

2.1.2 Open learning environment

Communicating through a screen can also create a sense that communication in the training session is only one way. As a result learners can feel like questions and discussions are not welcome. Learners can feel uncomfortable sharing questions in class due to the inability to personally and privately approach the facilitator (i.e. during break times), or being unable to determine an appropriate time to ask their questions as they feel that it will disturb the session. They can feel rushed to process training content during breaks and then struggle to form their questions clearly, or forget their questions during the Q&A portion if they did not have enough time to note it down. The development of this environment can keep learners from getting the most out of their training session.

2.1.3 Learner engagement

Maintaining learner engagement is a challenge for many educators both face-to-face and virtually. However, virtual learning naturally propagates disconnection and hinders the establishment and maintenance of learner engagement and participation. A big contributor to this is the term known as "Zoom fatigue" – a new term coined as being the sense of tiredness people feel after attending multiple live meetings throughout the day. The prospect of 'tuning' into full multiple days of training sessions can lead to personnel "switching off."

The lack of facilitator presence (learner-facilitator), the physical and mental isolation (learner-learner) that comes with virtual learning, limits the ability for learners to actively collaborate in group settings, discuss complex concepts and complete activities with 'live' and 'adaptive' guidance from the facilitator. The challenge is to develop alternative training materials that can effectively achieve the learning outcomes, and are as engaging [as when delivered face-to-face] and suitable for delivery via commercial communication platforms (learner-content).

2.1.4 Self-regulated learning and motivation

Educators have long worked to motivate students to want to learn. With the transition to virtual training sessions, encouraging learners to stay self-regulated and remain motivated with limited external support continues to be a challenge.

Virtual training sessions lack the collegiate structure traditional face-to-face training provides, inciting distractions and diminishing productivity. The lack of internal motivators (i.e. curiosity, interest, self-fulfilment) can mean learners are easily distracted when participating in a training session at home. Common distractions include social media (e.g. Facebook, texting friends or family), work (e.g. emails from managers, tight deliverables) and family (e.g. disruptions while learning from home). With the isolation and inability of the facilitator to monitor learner activities through the screen, distractions are much more challenging to manage and prevent in a virtual setting.

In traditional face-to-face training, there is an underlying learner-learner conformity – a learner is less likely to be distracted if the rest of the group is paying attention and completing the required activities. Even with virtual group activities, learner-learner disconnection is still prominent because it is much more challenging to form connections and build rapport online. Due to this, a sense of group accountability may be lacking online which can often result in some people doing all the group work online. As a result, their counterparts do not end up achieving the same level of concept understanding and application because of their lack of participation. There is a need to make up for this lack of connection when planning and designing virtual training sessions.

2.2 Working with technology

Various technology platforms (e.g. Zoom, Microsoft Teams, WebEx) have been incorporated into workplaces to overcome the inability to conduct face-to-face training sessions. Although technology can bring many benefits to training sessions, it is important businesses are aware of the technical limitations which can arise, have enough computer literacy to conduct sessions and are selecting appropriate technology to use.

2.2.1 Technical difficulties

In a virtual learning environment, technical difficulties are inevitable given the heavy dependence it has on learners having access to a stable internet connection at a sufficient speed. It can range from screensharing or videos not loading, internet connection constantly dropping out, to computers not being able to connect to the platform. Communication or exercise platforms must also be compatible with other required software (e.g. Apple vs. Microsoft programs) and client privacy requirements to ensure learners are able to attend the training session. If technical errors occur, learners may not be able to communicate with the facilitator, effectively understand the training content or have access to the training materials. This would not only result in lost time, as a result of facilitators having to resolve the technical difficulty, but also frustrates and disrupts learners – ultimately resulting in reduced focus and performance, as well as discouraging learning.

2.2.2 Computer literacy

It is also important to note, just as technical difficulties can diminish facilitator and learner motivation and drive, so can the usability of a software or program. Both facilitators and learners must have a basic degree of technological proficiency to be able to use the platform effectively and be aware that some platforms might not be compatible or suitable for the client.

Facilitators will need to know not only how to use the technology incorporated in the training session, but also understand enough to troubleshoot any technical difficulties. It is therefore crucial: (1) facilitators familiarize themselves with the software, (2) be aware of updates and changes to the software and, (3) understand their audience (i.e. learner age group, proficiency in software, exposure to virtual training sessions) in order to effectively communicate how to troubleshoot and resolve technological issues and minimize the disruption they introduce.

2.2.3 Cost benefit

It is important facilitators evaluate the cost benefit of investing in a platform or software and understand the limitations of their company when digitizing their traditional face-to-face training to a virtual setting. Different platforms offer different levels of compatibility, usability and accessibility that needs to be measured against a facilitator's needs, budget and training delivery.

For example, a company with technological experience may have the resources and skills to develop their own virtual learning software with increased security settings, customized tools and easy user interfaces that can be optimized to their needs and preferences. However, for most companies that do not have a heavy focus on providing training or education, this is not a reasonable investment and so these companies are limited to commercially available platforms. The challenge comes in assessing the suitability of these commercial platforms for use by both the facilitators and learners. Facilitators are challenged to think ahead and ensure platforms are accessible to learners. Further, different facilitator methods of conducting training, from the flow of a presentation to the types of exercises they develop have a cost impact. This needs to be weighed against the facilitator's requirements for training (does it allow you to design your exercises and activities as you envisioned for the learners to attain their learning outcomes? Can another software platform or application provide the same services/features for a lower cost?).

3 Planning to conquer the challenges of a virtual environment ('the good')

The key to successful delivery of training in a virtual environment is to undertake thorough planning. In the plan, do and review approach, the majority of time is usually spent in the planning/preparation phase. An increased focus in this area can enable the facilitator to pre-empt what can lead to poor training outcomes ('the bad') and prepare solutions ('the good') to overcome them if they arise. Sherpa has developed a 12-step approach for developing successful virtual training sessions. The planning and preparation stage make up two-thirds of this approach and consist of 8 key steps, as displayed in **Error! Reference source not found.**.

Figure 3.1: 12-steps to developing successful training sessions

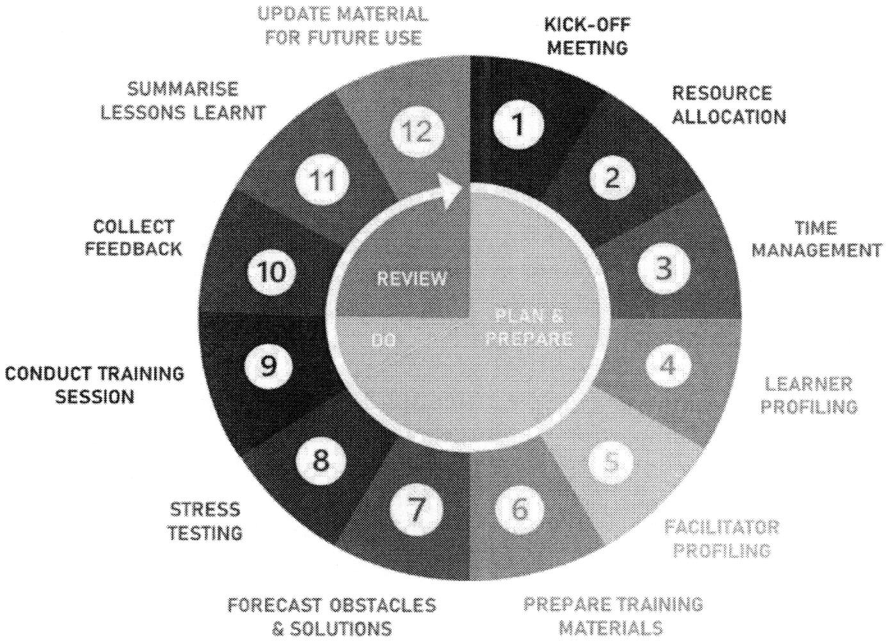

Completing these 8 steps can provide the facilitator with operational agility. This section will focus on Sherpa's experience and lessons learnt in the following key steps of the plan and prepare stage: learner profiling (#4), facilitator profiling (#5), preparation of training materials (#6), and stress testing (#8). These steps will demonstrate how Sherpa applies thorough planning to overcome the key challenges outlined in Section 2 and successfully provide virtual training sessions.

3.1 Learner profiling (#4)

Learner profiling is an effective way to better understand your audience (learners) and tailor the training session to better suit their needs. Information such as the learners' position or role in a facility and their level of familiarity and understanding of the topics in the training session (e.g. familiarity with the software in software training), provides useful insight for developing personalized training sessions. This information can help create industry specific exercises and examples that can be better understood by the learners. Providing learners with a quiz to complete prior to the training session can be a good way for the facilitator to gain insight into the learners' level of understanding, enabling them to adjust the session.

3.2 Facilitator profiling (#5)

Along with a learner's technical capabilities, it is important to understand the style of teaching required by the learners. Profiling facilitators can help to identify which facilitator will be able to best deliver the training to improve learner experience. When providing training to ESL learners, more time may be required to overcome communication barriers. To resolve this, Sherpa selects facilitators well suited in their pace of delivery (ability to translate complex concepts to basic English), as well as those who can best understand the learners (e.g. accents, understand common areas of difficulty and able to answer questions well) when planning for the session.

3.3 Preparing training materials (#6)

Step #6 is centred around training resource development. Before developing materials, facilitators need to decide on their delivery style and content structure as this will impact how resources will be prepared.

3.3.1 Harnessing the power of different communication platforms

When developing training sessions, it is best to select the platform early in the development process so facilitators can tailor their training sessions to the platform. From Sherpa's experience the choice of delivery platform can be heavily influenced by the client's company restrictions and by the functionality required.

For some clients and learners there are restrictions on which communication platforms they are permitted to use. It is important to discuss with the client to understand if any of their company restrictions will impact the training session. In some cases the client may not have restrictions on the platform but rather a preference based on what is most familiar to their team. Although learner familiarity to a platform might not seem as if it should influence the choice as much as functionality or company restrictions, it can make the learners feel more comfortable during the training session as well as remove the need to introduce how to use the platform. These too can play a part in improving the learner experience and should be considered.

On the surface different communication platforms may seem to provide users with the same basic functionalities, however, there are some subtle differences between them. When developing a virtual training session it is important to identify what functionality (e.g. multiple screen sharing) is required for the training so the best platform can be selected. When developing training courses for clients in the past Sherpa has made different choices to suit the type of training to be provided. For training sessions which heavily focused on learners completing worked examples and activities to reinforce learning outcomes, a multi screen sharing platform was selected. A multiple screen sharing platform such as Zoom allowed facilitators to watch over learners completing activities during the training session and provide assistance as required.

3.3.2 Recognizing technology is an enabler and not an end to itself

Part of developing virtual training sessions is the need to develop activities that can be completed in a virtual environment. It is important facilitators understand technology serves as an enabler as opposed to being the focus of the session. Some facilitators can be caught in the trap of getting lost in selecting software or applications based on the one with the most features to incorporate into their training sessions, when they do not benefit the learners.

The best place to start when deciding what software or application will best suit an activity is to understand the basic features of the activity and what it aims to teach the learners. When this is known, an application can be selected to ensure the activity still achieves its aim. In some cases selecting a basic user-friendly application can best suit activities. Sherpa has converted face-to-face activities to a virtual environment by using an application as simple as Microsoft PowerPoint. These activities required learners to form small groups and rearrange printouts ('drag and drop'). Although there are many applications which can provide this functionality, PowerPoint was selected as it is easy to use and familiar to many people. Particularly when developing activities for training sessions to be delivered to ESL learners, it is important to ensure instructions for activities can be simple. Selecting an application such as PowerPoint enabled learners to focus on developing their skills and not waste time learning how to use the application.

With the transition into a virtual environment, it has become simpler to incorporate quizzes into the training session. These can act to both keep learners engaged in the training session and provide facilitators with feedback on which areas might need to be reviewed. Applications such as Microsoft Forms has enabled the creation of short quizzes which can be incorporated into training sessions. These quizzes have the advantage of being able to instantly generate results which can be viewed by the facilitator and then used to shape the training session. Although this does not replace the information a facilitator receives from reading the faces of learners in a training session, this can help provide some insight into learner understanding.

3.3.3 Session pacing and structure

When developing a training session, it is important to consider how the content and session will be received by the learners. For this reason, it is particularly important in a virtual environment to incorporate interactive segments, regular breaks and build an open learning environment which welcomes questions and discussions.

Interactive segments are a great way to keep learners engaged as well as provide facilitators with an opportunity to receive feedback on content which might need to be re-explained. Sherpa has incorporated this into their sessions by selecting a learner to answer a multiple-choice question to review content. Questions can also be asked in the form of polls. This provides facilitators with feedback and is a great way to assess if the training session has been moving at an appropriate pace and if the explanations they have provided have been understood. Facilitators can maintain learner engagement by informing them they will be quizzed throughout the session as they will be more likely to pay attention to the content presented.

When training sessions are broken down into smaller sections with regular breaks, learners are more inclined to focus. Short regular breaks also offer learners the opportunity to ask questions and highlight misunderstood concepts to the facilitator. This allows the facilitator to better understand areas which might require clarification and more attention in future training sessions.

For learners to be comfortable with asking questions, it may be helpful to advise them on the best way to ask questions (i.e. via a chat, by raising their hands, etc.) and encourage them to do so. In order to build rapport, facilitators should provide learners with the opportunities to participate in discussions (e.g. introductions, safety moments). Facilitators should also be mindful of the way they provide feedback or answers to learners to not discourage them.

3.4 Stress testing (#8)

With the use of technology comes the potential for connectivity issues. As the mode of delivery in a virtual environment is prone to these connectivity issues it is critical more effort be spent on minimising the chance of failure. Sherpa has adopted 'testing sessions' with clients a week before training sessions to minimise potential connectivity issues that could occur at the beginning of a training session. These sessions ensure all learners can successfully connect to the session and equipment (e.g. headsets, speakers, microphones, screens, etc.) is appropriate and compatible with the chosen platform. This testing session is particularly important when using a platform unfamiliar to the client, such as Zoom, as it also provides the client with an opportunity to familiarise themselves with joining the session and using the platform before the training session. By allowing learners to familiarise themselves with the platform, this can remove the need to spend time during the training session to introduce it and allow the training session to run smoothly.

4 Conclusion

As businesses adapt to the 'new normal' in response to the changes catalysed by COVID-19, workplaces have seen a shift in the way work activities are prepared, conducted and delivered. Recognizing critical business activities must find a way to continue in this uncertain time has driven the need to develop innovative solutions and ensure they can still be done. A key area impacted by these changes is process safety training. In a world where face-to-face training is not always possible, facilitators have had to develop ways to continue this critical activity. Sherpa has recognized that by taking time to plan and prepare how to correctly select and apply technology to training sessions, businesses can begin to bridge the gap between face-to-face and virtual training.

Sherpa's experiences in developing virtual training sessions has identified some of the challenges of the virtual environment. The disconnect between the learner and facilitator as a result of communicating through a screen make it difficult for facilitators to read non-verbal cues, create an open learning environment, maintain learner engagement and motivation. Along with these issues, there are some challenges that arise from the use of technology such as technical difficulties, requirement for facilitators and learners to have computer literacy and analysing the cost benefit of various software and applications. These challenges pose a risk to the effectiveness of virtual training courses and as a result companies should take time to develop innovative solutions to overcome them.

It is important to understand each company's approach to virtual training will differ based on their own limitations. However, each company can apply the plan and prepare, do, review steps to their own situations to develop solutions with the resources available to them. Sherpa has applied the 12 steps of this process and has identified solutions to equip facilitators with operational agility in a virtual environment by: a) harnessing the power of different communication platforms; b) recognizing technology is an enabler and not an end to itself; c) preparation and stress testing; d) early profiling of learner and facilitator abilities; and e) the creation of a flexible teaching environment whereby the training modules are emphasized with the use of visual prompts, examples and activities to solidify concepts. In applying this approach, Sherpa has been able to successfully develop and deliver virtual training sessions to national and international clients with positive feedback and results.

SPRING21
+17TH GCPS
A Joint AIChE and CCPS Meeting

Virtual DHAs – Lessons from the Kitchen Table

Michelle R. Murphy
Mica, LLC
New Hampshire, USA
murphy@micadhm.com

Prepared for Presentation at
American Institute of Chemical Engineers
2021 Spring Meeting and 17th Global Congress on Process Safety
Virtual
April 18 - 23, 2021

Virtual DHAs – Lessons from the Kitchen Table

Michelle R. Murphy
Mica, LLC
New Hampshire, USA
murphy@micadhm.com

Keywords: dust hazard analysis, DHA, virtual DHA

Abstract

The 2020 pandemic required everyone to shift how they approached and implemented their work. Kitchen tables became desks, TVs became monitors, conference rooms went digital, and effective communication became more important than ever. For those of us planning to lead DHAs on-site at facilities, we had to figure out how to do this virtually...feel our way around the unknown during the unknown. This abstract will present the challenges faced, nuances learned, and the approaches used to navigate both. DHA considerations to be shared: Preparation, Document Sharing, Team Engagement, Getting the Whole Story, and Schedule and Timing. Rethinking the status quo improves processes. While we were all just trying to keep our heads above water, we stumbled across improvements that are here to stay. The presenter will share the pros and cons of conducting DHAs virtually, the lessons learned along the way, and which strategies worked best to keep moving forward on all DHAs.

1 Introduction

In early 2020 I was already scheduling multiple DHAs across the country. I conducted a couple in January and February and was scheduling more throughout the remainder of the year. I started buying plane tickets and booking hotel rooms. The next one was scheduled for March 10-13 in Raleigh-Durham, North Carolina.

On March 2, after planning, preparation, and kick-off meetings, one week before my flight, I received notice that the client corporation was suspending all visits to their manufacturing facilities and the DHA was being postponed. A year later, the plane ticket money is still in my travel bank with JetBlue.

At that time, things got eerily quiet in consulting as manufacturers adapted to new PPE and social distance requirements to keep their people safe amidst a global pandemic. As weeks turned to months and people adjusted to pandemic protocols, it was time to figure out how to get back on track with the many safety initiatives conscientious companies undertake every year.

In June, my biggest client and I started discussing doing DHAs remotely.

2 Approach

Remote conduct of DHAs would require additional planning, coordination, and commitment from both DHA leaders and team members. Small DHAs, for example project DHAs, with a day or less of team meeting time had successfully been completed remotely. It had never been attempted for a full manufacturing facility that would normally require 5 days on-site of team meeting time.

Like any analysis, DHAs follow a typical pattern of preparation, analysis, and follow-up.

1. Initial meeting/discussion with facility personnel
2. Document delivery to team leader
3. Scope definition and scheduling
4. On-site visit (Opening Meeting and Team Meetings)
5. Follow-up and reporting

Additional activities, such as response to the DHA recommendations, take place after completion of the team meetings and are outside the scope of this paper.

After developing and utilizing this approach in the course of 10 years, adjustments needed to be made for a virtual DHA. Many challenges could be overcome by sitting in a room with people, looking them in the eye, reviewing documents, walking around the facility, and building consensus. Due to the communication challenges of virtual vs. in-person meetings, virtual DHAs required shifts in preparation, document sharing, team engagement, communication, and scheduling.

2.1 Preparation

Typical preparation activities involved an initial meeting/discussion, requesting and receiving documentation from the facility, scoping, timing, and scheduling the team meeting dates.

Without in-person meetings, preparation became even more important. The first task in preparation is to communicate with the facility to introduce the concept and implementation of the DHA, explain the documents that are needed to prepare for the study, review the DHA activities, and describe the personnel needed for the team meetings. The team meeting approach of a DHA is often a surprise to people participating in their initial DHA. It is important for the facility to understand and provide the right personnel for the team meetings. This communication can be completed in an Initial Meeting or Initial Email. Items to be included in either (or both):

1. Introduction of team leader and facility personnel
2. Brief explanation of DHA process
3. Document request (a questionnaire can be helpful)
4. DHA activities and timeline
5. Team meeting needs

As part of any DHA preparation, the team leader reviews drawings and other process information to make an initial estimate of the nodes to be generated during the team meetings.

One common approach to breaking the operation into sections (or nodes) is to separate into individual unit operations and review each operation against a checklist. For dust handling operations, this can include:

- Bulk unloading
- Silo storage
- Pneumatic conveying
- Filter receivers
- Use or surge bins
- Dust collectors
- Sifters and screens
- Bucket elevators
- Mechanical conveyors
- Mills
- Ovens and dryers
- Packaging units
- Big bag unloading
- Charging stations
- Buildings housing equipment

Each unit operation is reviewed in succession using a checklist specific to the type of equipment. The DHA approach is described in detail in Murphy and Borene [1]. Typically, the nodes are developed during the study, with the input of the team. This is often accompanied by a set of drawings that are highlighted and labelled to match the nodes.

To help streamline the team meeting discussions, an adjustment to this process was made for virtual studies. The nodes were generated in advance by the team leader and reviewed with a team in a Node Review Meeting:

1. Detailed generation of nodes as a first pass by the team leader
    a. Drawing numbers
    b. Equipment numbers
    c. Highlighted drawings
2. Node review meeting to review the first pass node list with facility personnel
    a. Confirm unit operations to include
    b. Document safe operating ranges and design intent
    c. Establish dust properties to apply
    d. Provide basis for tentative DHA schedule

One of the unforeseen benefits to this node review meeting was that personnel became familiar with the types of information that would be needed during the team meetings and were better prepared.

2.2 Document Sharing

In a DHA (under normal circumstances) the team meets in a conference room, walks the plant together, reviews drawings together, and may tape them up on the wall for all to see. Whiteboards are used to detail specific operations, as needed. Adjustments to this process for virtual studies required some new tools:

1. Video conferencing platform
2. Screen sharing
3. Adobe commenting
4. Pictures
5. Microsoft whiteboard

While video conferencing software is not new, it was typically not used to host full day meetings of people scattered in multiple locations. It had been successfully used software for small DHAs and now it was going to be used for full studies. In a DHA, a textual recording of the discussion is projected onto a screen for the team to form a consensus. Sharing screens using video conferencing is standard practice and a simple adaptation.

Sharing process drawings was a separate challenge. How were drawings going to be highlighted and shared in real time? They could be highlighted ahead of time and emailed to the team. Two challenges with this approach included updating and re-sharing and flipping screens back and forth.

Anyone who has been in a DHA knows that sections are fluid. Often, drawings are not up-to-date and need some markup. Once highlighted and scanned, it would be difficult and time consuming to update them and re-scan to share. Team leaders are not experts in the operating process, any initial section breakdown and process highlighting would need adjustment when the whole team was assembled.

Another challenge was how to keep people focused when flipping screens between the DHA record and highlighted drawings. It can be difficult to follow along when someone else is controlling the flow of a document, even more so when flipping back and forth between multiple documents.

This challenge was overcome using a pdf software with commenting capability. The free version of Adobe Acrobat, Adobe Acrobat Reader DC, has a commenting function that allows marking up drawings on-screen. The drawings were highlighted during the preparation phase and shared and marked-up (both process changes and highlights) live during the team meetings. This allowed the team members to view both the drawing and the record at the same time, streamlining the process considerably. Figure 1 presents an example screen share during a live DHA session.

Pictures became one of the most important parts of the DHA. In a DHA, a walk-through of operations is conducted, notes are written, and occasionally a picture taken. Walk-throughs are a core part of a DHA. During this walk-through, current design, installation, and operation are spot-checked for electrical equipment classification, bonding and grounding, dust-tightness of equipment, fugitive dust build-up, etc. Communication with facility personnel is a key component during this walk-through to build trust, share knowledge, and identify areas that need immediate attention. When the team leader is an outside consultant, pictures are typically not allowed. Occasionally, an employee may take pictures during the walk-through but standard practice is to rely on notes and communications from the walk-through.

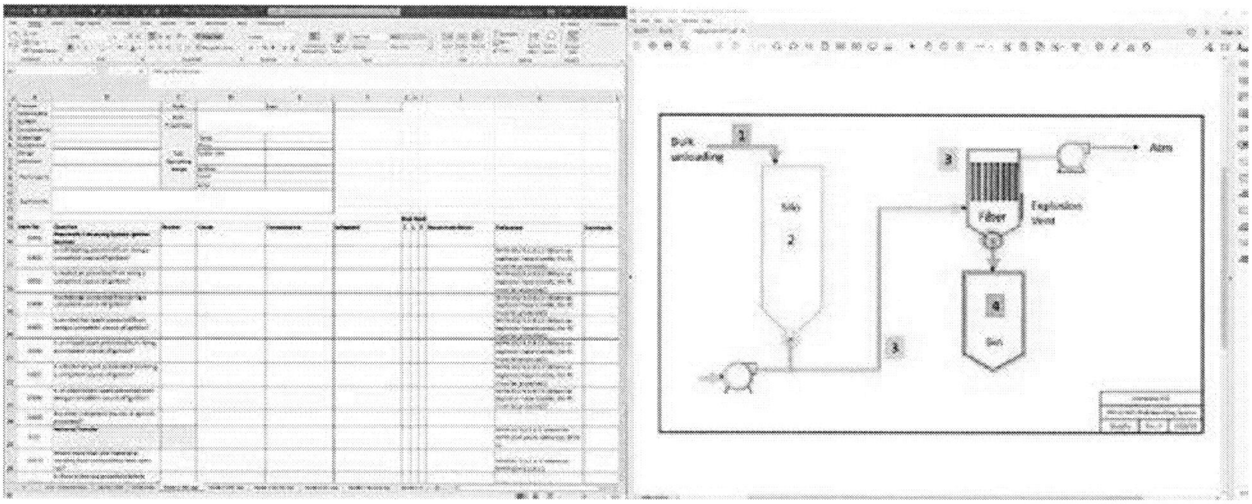

Figure 1. Screen share during live DHA sessions

In addition to drawings and pictures, another tool would be missing from a virtual DHA, a whiteboard. In any team meeting, there comes a time where people need to visualize what someone is describing. To show updates not included on a drawing, better explain the operation of a piece of equipment, or convey other operational details, whiteboards are utilized throughout DHAs. This missing piece was replaced with Microsoft Whiteboard©. An example sketch to convey the difference in design of a filter receiver vs. a bin vent on a bin is included in Figure 2.

2.3 Team Engagement

Another major challenge was keeping the team involved and engaged. DHA team meetings are conducted on-site with a team of people sitting in a conference room, all accountable to one another. While people get distracted with email, phone calls, and other disruptions, they are generally kept engaged by their physical presence in the room.

In virtual DHAs, everyone may be in a different room. It is much easier for people to get distracted by the business of the day and stop participating. Two ways to overcome this include having facility people gather in a conference room within the rules of current COVID-19 protocols and requiring people to keep their cameras on. The first approach was utilized successfully at several facilities and certainly streamlined communication and supported the overall goal of the DHA. The second approach, requiring people to keep their cameras on, in theory, would force people to stay more engaged. In practice, it can have very negative effects. Many people were forced into working from home with no warning and little control over their surroundings. People may not have had the opportunity to setup and prepare a suitable workspace for sharing on camera. The needs of other individuals in the home may also create noise and distractions that people do not want to share with their co-workers. Cameras make people more self-conscious and on edge leading to increased stress and anxiety. Forcing people to overlap their personal life with their professional one was not the role of such a team meeting. An article by Jeremy N. Bailenson [2] frames a number of issues leading to psychological consequences and fatigue from videoconferencing.

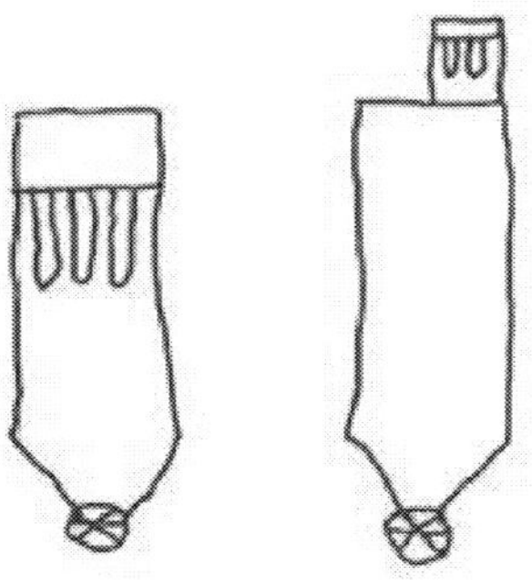

Figure 2. Whiteboard sketch depicting a filter receiver and a bin vent

In one study, we had all of the plant personnel in a conference room. There they were able to share drawings and other process information to reach consensus much faster than if they had all been in separate locations.

With respect to cameras, a compromise was made. Cameras were requested during a short opening meeting so that people could place names, faces, and voices for continued collaboration throughout the rest of the team meetings. In the end, it was found that the best way to keep people involved and engaged was no different than face-to-face meetings: keep the discussion moving, call on people by name when information is needed, and listen to others.

2.4 Getting the Whole Story

The biggest challenge with a virtual DHA is that we are relying on the perception of people who work in the plant every day. In the CSB's Call to Action: Combustible Dust [3] they noted:

> "Didion employees had varying perceptions of dust accumulations within the facility."
>
> "Some employees characterized the plant as 'spic and span,' while others reported that dust was constantly present in the work area."

Figure 3 is a photo from that report depicting differences in perception. Each color presents comments from the same incident. This difference in perception is quite apparent.

Knowing that 'perception is in the eye of the beholder' it was imperative to find a way to bring everyone's perception to the same place. While on-site in a DHA, a lot of time is spent showing people dust accumulations and talking about levels. Demonstrating how little dust can obscure the color underneath, how elevated horizontal surfaces such as I-beams and pipe racks can be more than 5% of the floor space, or how fugitive dust contains a distribution of particle sizes likely to contain an explosible concentration if blown into a cloud is easily communicated during a walk-through. In a virtual DHA photos of building spaces were used to supplement discussions and demonstrate these same concepts.

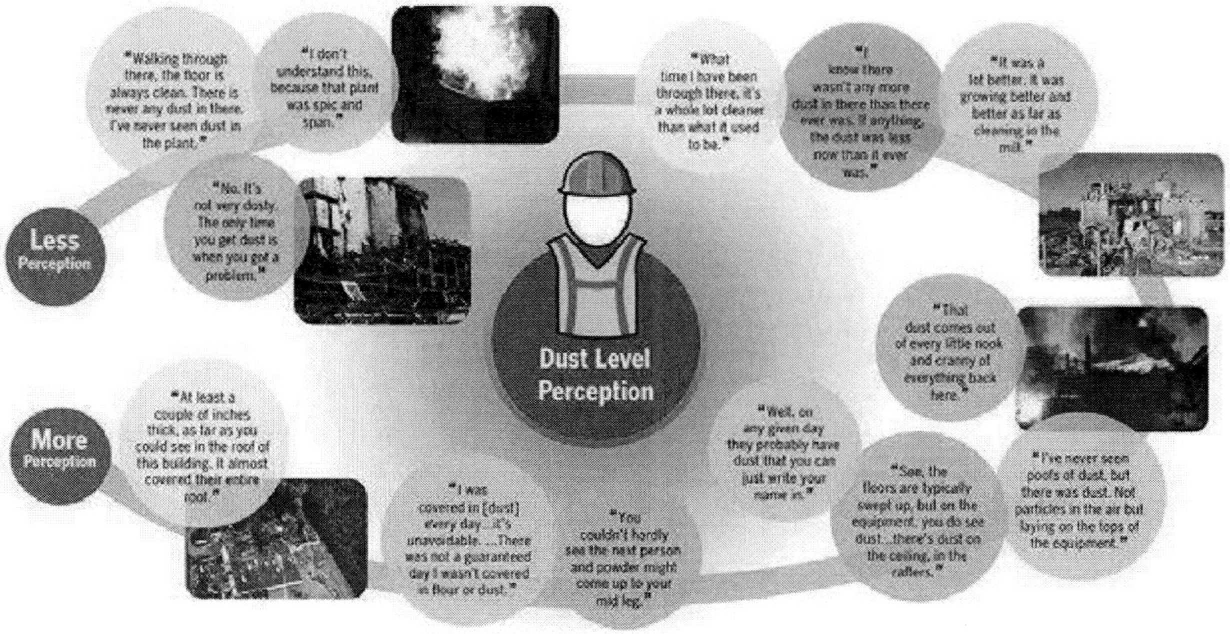

Figure 3. Excerpts of employee perceptions of dust levels at Didion and other facilities from CSB's Call to Action: Combustible Dust

2.5 Schedule and Timing

A DHA of an average manufacturing facility can typically take 3-5 days of in-person team meetings. Often travel is involved so the meetings are held all day, on successive days, to be most cost-effective.

Virtual DHAs provided an opportunity to re-think this grueling schedule with many options. Several models were attempted:

- Non-consecutive days spanning multiple weeks
- Four days per week spanning multiple weeks
- Shorter hours over five days per week

There were pros and cons to all of these options and the schedule often more directly reflected the length of the DHA rather than a desire to spread the work over a higher number of days. For example, while non-consecutive days spanning multiple weeks is a less grueling schedule, it ties people up for more weeks, leaving fewer weeks to schedule other work.

Virtual DHAs also required more time for preparation and team meetings. Experience showed that 2 extra days of preparation time and 25% more team meeting time was needed.

3 Moving Forward

After conducting many studies virtually over the past year, the question becomes, as we get back to normal, do we go back to on-site studies or stay with virtual studies. In making these decisions we consider the following aspects of both:

1. Cost
2. Time
3. Effectiveness

For purposes of this paper, cost is figured as the cost of hiring an outside consultant to lead the study. The difference in cost of on-site vs. virtual was negligible in that the cost of the additional team leader level of effort was offset by the reduction in travel costs. This was not a differentiator.

Aside from the additional preparation time encompassed in the cost described above, about 25% more time is needed for team discussions. This is due in large part to the time involved in sharing photos and lengthier discussions in reaching consensus.

The definition of hazard evaluation can help to measure effectiveness of on-site vs. virtual DHAs. CCPS defines hazard evaluation in their book, Guidelines for Hazard Evaluation Procedures [5], as:

"A *hazard evaluation* is an organized effort to identify and analyze the significance of hazardous situations associated with a process or activity."

NFPA in their standard on the fundamentals of combustible dust [6], states "The DHA shall evaluate the fire, deflagration, and explosion hazards and provide recommendations to manage the hazards…"

In evaluating the results of the DHA, did we 'identify and analyze the significance of hazardous situations'? Did we 'evaluate the fire, deflagration, and explosion hazards and provide recommendations'? By all accounts those goals were accomplished. Each unit operation was evaluated for hazardous situations, existing safeguards were identified and evaluated for adequacy, and recommendations were made where deficiencies were noted.

Is it possible to conduct DHAs virtually? The answer is yes. Should this be standard practice moving forward? The answer is no.

While the virtual DHA was an adequate stop-gap measure during a pandemic, experience suggests that while overcoming individual perceptions of hazards is improved with the use of pictures, it is not removed altogether.

It would be beneficial to evaluate the true effectiveness of a virtual vs. on-site DHA using a standard metric. However, the dependency of the effectiveness of a DHA on the team members assembled makes it almost impossible to compare one to another.

3.1 What to Keep

While 'necessity is the mother of invention' many of these tools also improve, streamline, and benefit on-site DHAs. These include:

1. Additional preparation, including a Node Review Meeting
2. Improved document sharing, including live on-screen highlighting of drawings and use of pictures

The Node Review Meeting provides the benefit of preparing the team for the activities to come during the team meetings and gives some additional time for the facility to chase down the necessary information, such as operating procedures and dust characteristics. It was found to streamline and shorten the length of the team meetings.

Having pictures was extremely helpful in resolving differences of opinion, answering questions, and most-importantly – keeping people in their seats. These benefits would also be applicable to on-site DHAs.

3.2 What Not to Keep

The virtual nature of virtual DHAs. As stated above, it is this author's opinion that better overall quality is found with an on-site approach.

4 References

[1] M.R. Murphy and M.J. Borene "How to Conduct a Dust Hazards Analysis," *Process Safety Progress*, March 2018.

[2] Jeremy N. Bailenson "Nonverbal overload: A theoretical argument for the causes of Zoom fatigue," *Technology, Mind, and Behavior*, Volume 2, Issue 1, 2/23/2021. Available at https://tmb.apaopen.org/pub/nonverbal-overload/release/1. Accessed on March 11, 2021.

[3] CSB Call to Action: Combustible Dust Didion Milling and Others, Available at https://www.csb.gov/assets/1/6/call_to_action_-_final1.pdf. Accessed on February 15, 2020.

[4] CCPS. *Guidelines for Combustible Dust Hazard Analysis.* Center for Chemical Process Safety, American Institute of Chemical Engineers, New York, NY, 2017.

[5] CCPS. *Guidelines for Hazard Evaluation Procedures.* 3rd Edition. Center for Chemical Process Safety, American Institute of Chemical Engineers, New York, NY, 2008.

[6] NFPA. *Standard on the Fundamentals of Combustible Dust.* NFPA 652, National Fire Protection Association, Quincy, MA, 2019 Edition.

620242 Leading by Example – Innovations in Pharmaceutical Development to Position for a Sustainable Future

Abstract:

As the active substances and modalities for advanced therapies have become more structurally complex and R&D timelines continue to be compressed, the chemistry and technologies necessary to provide safe, robust and economical access to these molecules needs to keep pace. Enabling technologies such as continuous processing, automation and predictive modeling through first-principles, as well as machine learning can expedite process development, scale-up and technology transfer while ensuring enhanced process efficiency and delivering quality medicines to patients more effectively. The importance of strong leadership and novel concepts in pharmaceutical industry has become even more evident in recent Covid-19 pandemic crisis, different organizations are actively thinking of implementing these paradigms to their benefit. This presentation will share some of the leadership experiences and novel technologies that can be leveraged to accelerate process development and manufacturing and discuss the challenges and opportunities to position for a sustainable future.

620403 Enabling Innovation in New Modalities Synthesis

Abstract:

Dr. Cruz will speak on a subject related to leadership in the pharmaceutical industry.

620244 Getting Ahead of the Curve – Emerging Trends and Success Factors in Biopharmaceutical Development

Abstract:

This talk will discuss leadership aspects involved in pharmaceutical development.

618433 Imposter Syndrome

Abstract:

Imposter syndrome (IS) is a mindset wherein accomplished individuals feel that they have succeeded only due to luck or because others have not realized that they are frauds and thus not deserving of their success. This syndrome is an emotional response rather than a logical one in that the individual dealing with IS knows all details of his/her specific difficulties and struggles and therefore feels that they are undeserving of their achievements. IS is widespread and is characteristic of individuals irrespective of their experience level, gender, ethnicity, stature, and chosen field. IS causes us to avoid applying for certain jobs, negotiating effectively, and enjoying our accomplishments. In this talk, possible reasons that IS is prevalent will be described. Approaches to overcome this syndrome and to minimize the effects of IS on our careers, goals, and subordinates will be discussed

MBBR Wastewater Treatment System for Nitrification of Combined Industrial and Municipal Wastewaters under Cold Weather Conditions with Wide Load Fluctuations

Siva Angappan[1], Eugene Townsley[2] and Somnath Basu[1],

1. Headworks International, 11000 Brittmoore Drive, Houston, TX 77041
2. Batesville Wastewater Treatment Plant, Batesville, AR 72501

Abstract

Uncontrolled discharge of ammonia in treated effluent from wastewater utilities is harmful to the receiving bodies of water. At high concentrations ammonia is toxic to aquatic life. It also adversely impacts the presence of dissolved oxygen which is critical to the health of water bodies. Additionally, discharge of nitrogen in treated wastewater in all forms, including ammonia, promotes undesirable growth of aquatic plants e.g., hyacinth, algae, etc., also called eutrophication, which renders water unsuitable for use and recreational purposes. Environmental regulators throughout North America and globally are increasingly calling for effluent nitrogen control, which requires nitrification as the first step. Nitrification involves biological oxidation of ammonium ions in wastewater. It is kinetically a very slow process and sensitive to environmental conditions, especially wastewater temperature, pH, presence of heavy metals and other toxicants. Many treatment plants, especially in rural US and Canada are lagoon based that can remove only organic pollutants - BOD (biological oxygen demand). In winter, these lagoons are exposed to very low ambient temperatures. Nitrification is difficult to sustain under cold temperatures <5°C. Moving Bed Biofilm Reactors (MBBR) have been shown to effectively conduct nitrification under sustained low temperature conditions while consistently achieving low ammonia limits. Other benefit of the MBBR process include – (a) automatic growth and retraction of the biofilm under varying flows and loads without having the operator to adjust, monitor, or control parameters such as recycled activated sludge (RAS) and mixed-liquor suspended solids (MLSS), (b) requirement of much smaller reactor volume and footprint area, and (c) ability to quickly recover from hydraulic surges, and toxic and shock loads.

Batesville, AR, has a population of around 10,000 but its wastewater treatment plant treats 60,000 population equivalent flows due to many industrial facilities in the area including poultry and dairy industries. Prior to 2015, lagoons were used to treat the wastewater to remove only BOD. Lagoons were also used to settle the solids. Due to substantial growth in industrial discharges to the facility, the plant was stretched to the limit to remove BOD, especially in winter, when the water temperature in the lagoons reach as low as 4° C.

In 2015, the plant's permit was revised to remove ammonia-nitrogen with a year-around limit. Due to the very low water temperatures in winter expansion of the lagoons was not considered a viable option. MBBR was selected as the technology to meet the new regulatory requirements for BOD and ammonia-nitrogen control due to its ability to perform under very low temperatures and handle varying and toxic loads from industrial contributors.

This presentation will consist of an analysis of many years of operating data demonstrating how the MBBR process helped the treatment plant achieve its goals of producing stable effluent quality consistently over the year even under low temperatures and varying loads. Operating performance will be compared with kinetic models based on the principle of MBBR technology.

620741 Technical Women's Leadership Forum – What are we waiting for?

Abstract:

The Women's Leadership Forum (WLF) is one of the largest and oldest employee resource networks (ERN) at 3M. Yet until only recently, relatively few women at the bench (technical track) participated in this ERN. In addition, women reaching upper level technical positions such as Corporate Scientist or Division Staff Scientist, remained disappointingly low over the past several decades. In 2018, female scientists responded by creating a technical chapter (T-WLF) in order to specifically address the needs of technical women. Core WLF pillars of "I believe in Me", "I believe in You", and "We believe in Us" were reinterpreted to meet the needs of technical women. The journey began with the creation of deliberate efforts to 1- *Retain* existing technical employees with Career Development and Technical Men as Advocates committees, 2- Implement best practices and build a better pipeline with *Outreach* committee, and 3- Address the early gap for girls in STEM with *Inspire* committee. In short two years, team has accomplished to collaborate with executive management and amplified the voice of the female technical community, partnered with 3M's Office of Intellectual Property to conduct surveys and focus groups, successfully granted funding from 3M to two non-profit organizations, highlighted successful technical women stories on internal internet, increased presence and engagement with recruiting events, and initiated global expansion. Today, 3M technical women are not waiting; they are part of making change happen.

SPRING21
+17TH GCPS
A Joint AIChE and CCPS Meeting

The e-SafetyCase – Electronic or Effortless?

Gareth Ellor
Director - Clean Energy & Innovation
Risktec Solutions Ltd
Wilderspool Park
Greenall's Avenue
Warrington
WA4 6HL
UK
gareth.ellor@risktec.tuv.com

Prepared for Presentation at
American Institute of Chemical Engineers
2021 Spring Meeting and 17th Global Congress on Process Safety
Virtual
April 18 - 22, 2021

The e-SafetyCase – Electronic or Effortless?

Gareth Ellor
Director - Clean Energy & Innovation
Risktec Solutions Ltd
Wilderspool Park
Greenall's Avenue
Warrington
WA4 6HL
UK
gareth.ellor@risktec.tuv.com

Keywords: Electronic, Safety, Case, HSE, HSSE, ALARP, Risk, Management, Digitalization.

Abstract

An Electronic Safety Case (e-SafetyCase) is an alternative structured approach to safety. The purpose of an e-SafetyCase and the information contained within it remains the same as a traditional Safety Case; to demonstrate risks are reduced As Low As Reasonably Practicable (ALARP). However, instead of being a lengthy and complex document, the e-SafetyCase is highly interactive, using clickable links as a means of connecting and navigating the information quickly and easily. Not only does this make it simple, intuitive and user-friendly, but by integrating it with day-to-day operations, it is kept current and evolves with the facility. This ensures there is no disconnect between the safety case and operations; they become one and the same! As a result, the e-SafetyCase not only demonstrates risks are reduced ALARP but it ensures risks are managed ALARP throughout the facility lifecycle; the ultimate goal of any safety management system. If done properly, an effective e-SafetyCase will be so useful and helpful to the workforce that it will naturally evolve hand-in-hand with the facility as routine business. Effortless safety – is that not too good to be true?

This paper introduces the concept of an e-SafetyCase outlining why, despite being a concept which has been around for some time, it has struggled to gain widespread traction across industry. It debunks the myths about what an e-SafetyCase is, in an attempt to knock-down the barriers we consider are blocking its widespread use. The majority of the paper is devoted to demonstrating what an e-SafetyCase looks like using a practical case study, highlighting the significant benefits it delivers to the safe and reliable operation of the facility throughout its lifecycle. In addition, it demonstrates the flexibility and adaptability of the e-SafetyCase approach to allow the considerable benefits to be realized by any major hazard facility. Finally, it will touch on ground-breaking work being done to further improve the functionality and hence benefits of an e-SafetyCase by blending with wider technological innovations. For example, the integration of virtual and augmented reality to really bring the Safety Case to life through modelling of accident scenarios and emergency response arrangements. Although ambitious, this kind of forward thinking could ultimately change the landscape of safety management in major hazard industries.

1 Overview

This paper introduces the concept of an Electronic Safety Case (e-SafetyCase), outlining why, despite being an aspiration which has been around for some time, it has struggled to gain widespread traction across industry. It debunks the myths about what an e-SafetyCase is, to help knock-down the barriers blocking its widespread use. The majority of the paper is devoted to demonstrating what an e-SafetyCase looks like using a practical case study from an Offshore Wind Farm. This case study illustrates the significant benefits it can deliver to the safe and reliable operation of the facility throughout its lifecycle. In addition, it demonstrates the flexibility and adaptability of the e-SafetyCase approach to allow the considerable benefits to be realized by any major hazard facility. Finally, it touches on the ground-breaking work being done to further improve the functionality and hence benefits of an e-SafetyCase by blending with wider technological innovations. Although ambitious, this kind of forward thinking could ultimately change the landscape of safety management in major hazard industries.

2 What is a Safety Case?

Before we start to explore the e-Safety Case concept, we must firstly remind ourselves of what a Safety Case actually is.

A Safety Case is generally accepted as being the written demonstration that a plant, facility, operation, etc. has been designed, conceived and executed in such a way that the major Health, Safety, Security and Environmental (HSSE) risks incurred during its lifecycle are reduced to a level that is As Low As Reasonably Practicable (ALARP).

They are virtually always written documents and despite best intentions from some to break-down established precedents and norms, and perhaps a lazy approach from others, they often end up being lengthy and complex. Whilst serving their core purpose, they can easily become impenetrable to anyone without direct Safety Case experience, heavily limiting wider benefits.

They cover the major HSSE risks thereby ensuring that most effort is focused on managing the risks that could cause greatest impact. They rely on the wider HSSE Management Systems to effectively manage less significant hazards on a proportionate basis.

They are generally associated with a legislative regime and, if one were to be cynical, have evolved in such industries to be a document issued to a regulator to show risk levels are ALARP to obtain a licence to operate. However, in reality, a Safety Case is independent of regulations. It represents good practice. It shows all stakeholders that risk is being managed effectively.

Figure 1. Safety Cases - Dispelling the myths [Ref 1]

Figure 1 summarizes the myths we hear time and time again about what Safety Cases are perceived to be (on the left) with the reality of a good Safety Case on the right. This has prompted us to use the Marmite analogy to inject some humor into what is a very dry subject ("you either love it or hate it!"), to try and break down these myths and allow industry to embrace the Safety Case approach and realize the huge benefits it offers. People are rarely ambivalent to a Safety Case. They either love it or hate it. In our experience, the haters usually come with a skewed view of what a Safety Case is. Figure 1 shows the huge untapped potential of a Safety Case and the e-Safety Case is ready to deliver on this.

3 What is an e-SafetyCase?

The purpose of an e-SafetyCase, and the information contained within it, remains exactly the same as a traditional Safety Case; to demonstrate risks are reduced ALARP. It is how this information is presented that is the key difference. Instead of being a lengthy, complex written document, the e-SafetyCase is highly interactive and intuitive, using clickable links as a means of connecting and navigating the information quickly and easily. As a result, the e-SafetyCase is accessible and helpful to anyone involved in the safe operation of the facility.

The best way to understand what an e-SafetyCase is, and get an appreciation of the benefits it offers over a traditional paper-based Safety Case, is to take a look at a practical example.

We have developed an e-SafetyCase for an offshore wind farm being built in UK waters. Although not a regulated industry, the client showed a strong appetite for a Safety Case approach as they understood the benefits it delivers.

Before we explore this, we must firstly understand how it was developed as this has a significant bearing on how it is presented.

3.1 Risk Management Process

A Safety Case should represent the output of a proportionate and practical risk management process which is fully integrated with the project development process for a particular design. Figure 2 illustrates the process adopted as the foundation of this Offshore Wind e-SafetyCase and shows how this meets the three fundamental challenges of effective risk management (see Figure 3).

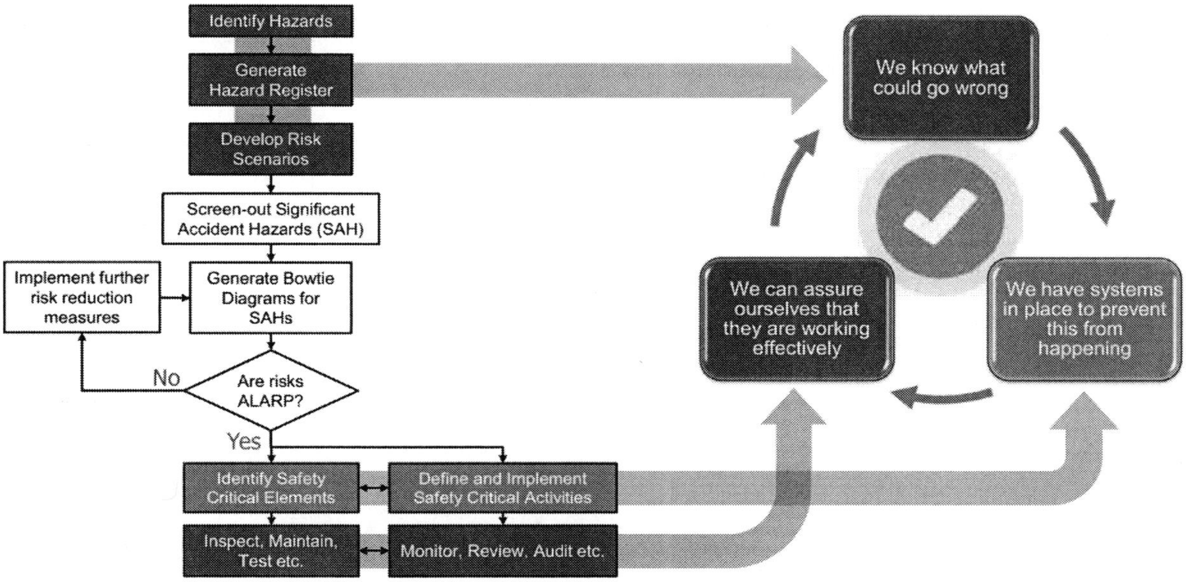

Figure 2. Risk Management Process – the foundation of the e-SafetyCase

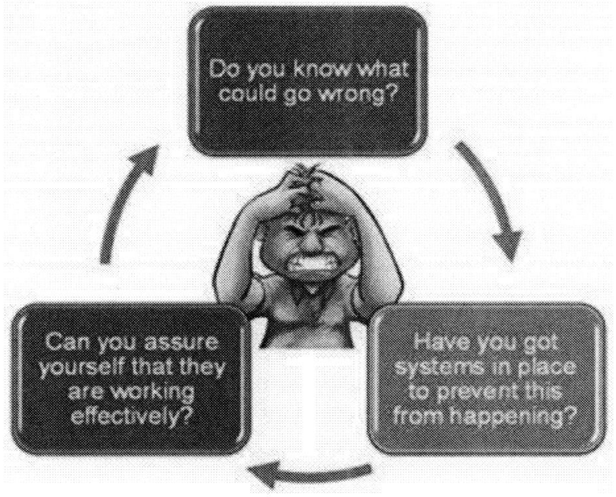

Figure 3. The challenges of effective risk management [Ref 2]

3.2 *Definitions*

Significant Accident Hazards (SAH) – Although not comparable to, say, the oil and gas industry in terms of ultimate consequences, if the risks associated with offshore wind are inadequately managed, they could still result in fatalities and loss of reputation. Whilst the term "significant" was used instead of "major" to define the hazards screened-out for detailed assessment, the approach is equally relevant to major hazard industries. The only difference being that in applying this approach to major hazard facilities, more hazards are likely to be identified for detailed analysis and/or the screening threshold may rise to ensure a proportionate approach.

Safety Critical Elements (SCE) are defined as those items of equipment or structures whose failure could cause or contribute to a SAH, or the purpose of which is to prevent or limit the consequences of a SAH.

Safety Critical Activities (SCA) are defined as those activities which are critical to ensuring safety and continued integrity of design, or ensure SCEs continue to function as intended.

3.3 *Bowtie Approach*

As illustrated in Figure 2 on te previous page, the Bowtie methodology was used to assess all SAHs.

The Bowtie technique is an extremely powerful way of clearly illustrating how risk is being managed within an operation, business, etc. It helps to ensure that risks are managed rather than just analyzed, partly by going beyond the usual risk assessment "snapshot" and highlighting links between the risk controls and management systems.

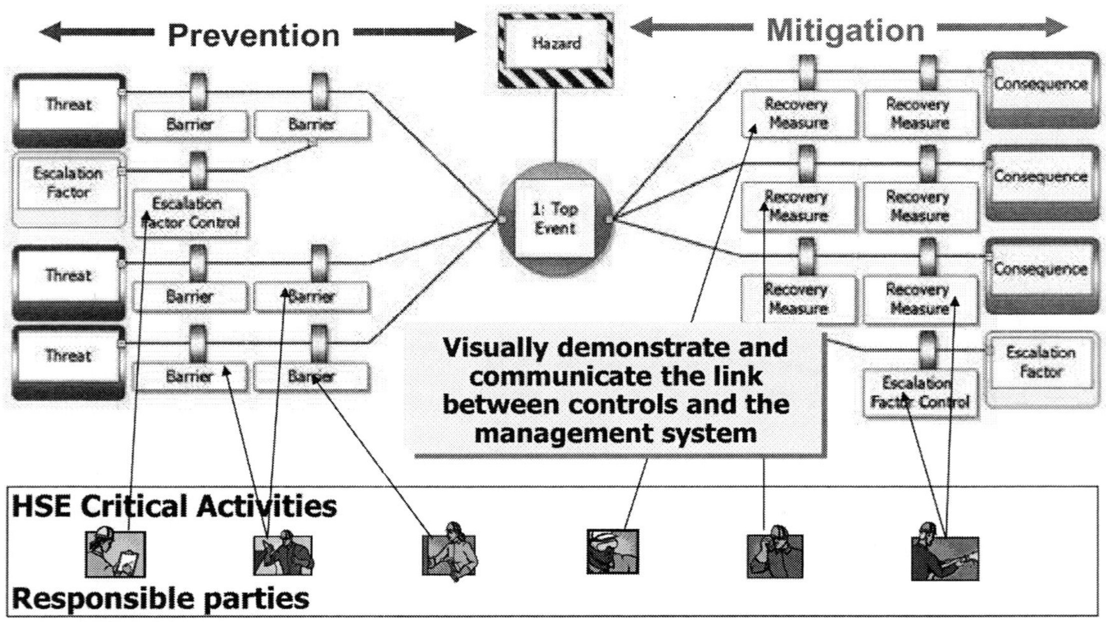

Figure 4. Bowtie Diagram

Figure 4 illustrates the basic structure of a Bowtie diagram, the key elements of which are:

- The **Hazard** which has the potential to give rise to the unwanted event you are concerned about. This is illustrated in the yellow/black striped box.
- The unwanted event you are concerned about if the hazard is "released". This is called the **Top Event** and is the red circle which forms the "knot" of the Bowtie.
- The credible causes of the unwanted even, termed **Threats,** and are illustrated in blue on the left hand side of the diagram.
- The unmitigated **Consequences** of the unwanted event. These are illustrated in red on the right hand side of the diagram.
- An **Escalation Factor** (or "degradation factor") which could compromise the integrity of a barrier.
- The **Barriers** in place to prevent the unwanted event occurring or minimize its consequences. Those on the left hand side of the diagram prevent the hazard being realized and are categorized as **Prevention**. Those on the right hand side limit the ultimate consequences of the hazard if realized and hence represent **Mitigation.** Barriers can also be included to eliminate or reduce the impact of an Escalation Factor on the integrity of a specific barrier, either prevention or mitigation.
- The specific systems, processes, procedures, etc. which must be in place to ensure the integrity of these barriers. This is achieved by identifying the Safety Critical Elements (SCE – plant, equipment, etc.) and their associated Safety Critical Activities (SCA – operations, inspections, checks, etc.), along with the roles, responsibilities and competencies of the people who ensure these work effectively.

This latter point represents the real benefit of the Bowtie methodology. It provides a clear, auditable trail from hazards right through to the specific systems, processes, procedures, etc. implemented, with performance standards and responsibilities clearly assigned, to reduce the associated risks ALARP. It therefore illustrates the complete safety assurance process in a clear picture which can be easily understood by everyone within the organisation.

3.4 Significant Accident Hazards & Bowties

The following Significant Accident Hazards (SAH) were identified for the offshore wind farm and Bowtie diagrams were generated for each of these:

1. Electrical hazards
2. Fire hazards
3. Loss of structural integrity
4. Personnel transfer
5. Personnel at height
6. Hydraulic and mechanical systems

7. Hydraulic systems – pressure hazards
8. Vessel transport
9. Helicopter transport
10. Suspended loads
11. External hazards
12. Explosion hazards

3.5 e-SafetyCase Overview

The e-SafetyCase we have developed is presented in a highly visual web-browser style. To illustrate just how simple this concept can be, the version we're going to show was developed in PowerPoint and turned into a .pdf file. It is therefore accessible on any computer and can be developed and updated using standard software on any PC. Furthermore, it is sized such that it can be readily saved on a laptop or tablet allowing it to be accessed by anyone, anywhere, anytime.

It can be navigated via two separate home pages which the user can toggle between.

Firstly the user can explore the e-SafetyCase from a process perspective (see Figure 5). This is based on a more simplified version of the risk management process illustrated in Figure 2. This allows the user to quickly and easily access the information based on the process through which it has been generated. This allows those familiar with the underlying risk management process to access the information they require in a very quick and highly intuitive way. It also brings the overall safety case together to show how all the elements combine to present a robust justification that risk is being managed ALARP. Finally, it raises awareness of the underlying risk management process to people less familiar with this, promoting an understanding of how risk is being managed across the business.

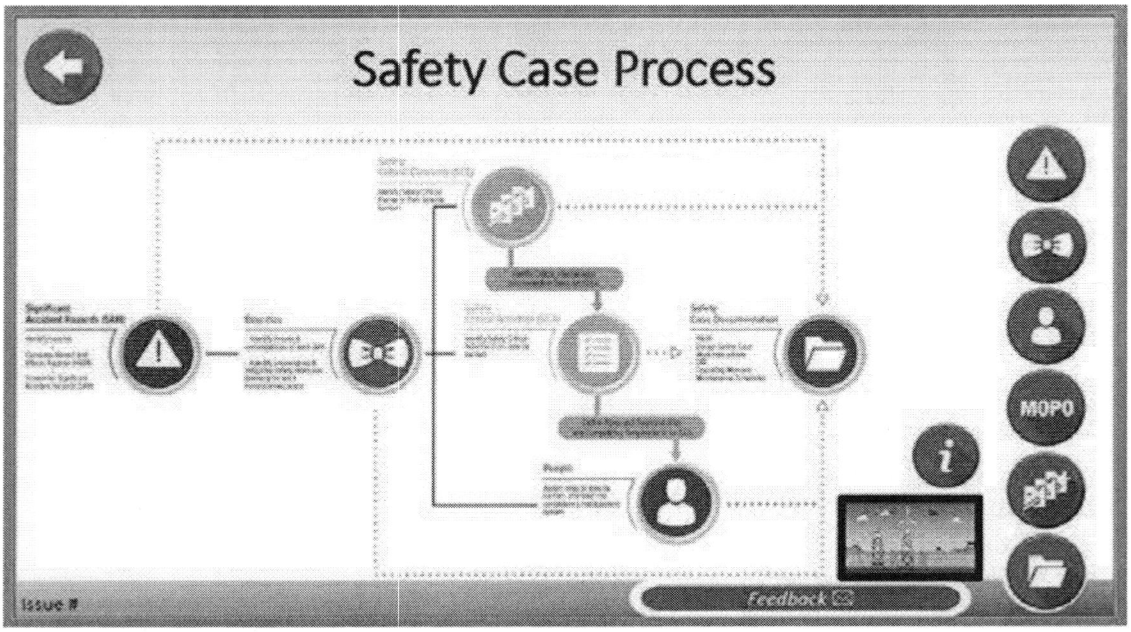

Figure 5. Process View Homepage

Alternatively, the user can switch to an infographic showing all the key elements of the overall wind farm to explore the e-SafetyCase from an asset perspective (see Figure 6). This is particularly useful to someone less familiar with the risk management process or perhaps wants to immediately dive into the detail behind how risk is being managed within a specific element (e.g. turbine, foundation, offshore substation, helicopter, etc.).

Figure 6. Asset View Homepage

On each homepage the user can click on the same set of buttons to access the following information:

Significant Accident Hazards (SAH)

Presents definition of a SAH, lists all SAHs and provides information on each individual SAH (description, video, training material, etc.).

Bowtie Diagrams

Lists all Bowtie diagrams relating to each SAH and allows user to view individual Bowties to varying levels of detail (overview to complete).

People

Identifies all safety critical roles claimed within the Bowties and assigns these to a responsible person, whilst listing all of the SCAs each person is responsible for.

Matrix Of Permitted Operations (MOPO)

An interactive tool to determine whether it is safe to undertake key operations, recognising the conditions which may elevate the associated risk and/or other simultaneous operations.

Safety Critical Elements

Lists all SCEs, rolled-up into groupings linked to the principle of hierarchy of risk reduction.

Documentation

Provides links to all associated Safety Case documentation (studies, analysis, etc.) which underpins the position of safety set-out within the e-SafetyCase. This represents the supporting evidence.

The "Feedback" button, when clicked, opens up an e-mail to allow the user to provide feedback directly to the e-SafetyCase owner (e.g. something unclear, something wrong, something which can be improved, etc.). The e-SafetyCase owner collates this feedback and folds this into periodic updates where appropriate.

Whilst the presentation of the Safety Case in such a simple, intuitive and user friendly format is a quantum leap from traditional paper based Safety Cases, the real beauty of the e-SafetyCase is that all of this information is connected together to allow the user to access it via whatever route (or "journey") they so choose, via quick and easy links. In very simple terms, the e-SafetyCase brings the Safety Case to life through the navigation triangle illustrated in Figure 7.

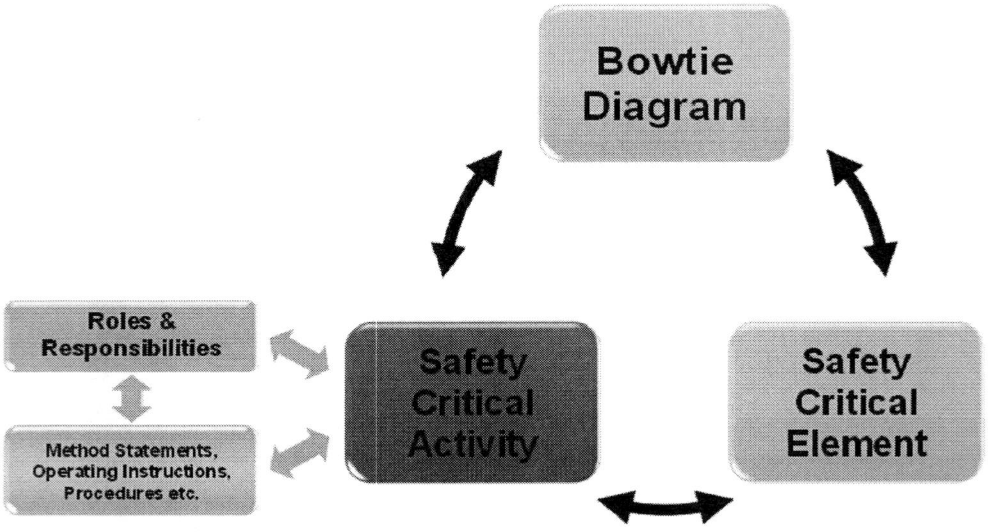

Figure 7. e-SafetyCase navigation triangle

For each SAH the user can, via single clicks, jump between the Bowtie, SCEs and SCAs from any entry point. This allows the user to:

- Gain an understanding of the role each SCE or SCA plays in ensuring risks are reduced ALARP by linking directly to the Bowtie and showing how each fits into the big picture.

- Link directly from a SCE to view the SCA(s) which must be performed to ensure that feature continues to perform its safety function correctly.
- Link directly from a SCA to the associated SCE to understand the feature the activity relates to.
- From a SCA, click directly to the method statement, operating instruction, procedure, etc. which must be followed to complete that activity correctly and identify who is responsible for completing that SCA.
- Quickly and easily identify all SCAs a specific individual is responsible for.

But this simple and easy navigation is not confined to individual SAHs. The links cut across all SAHs, Bowties, SCAs and SCEs and can be accessed both from a process and asset perspective.

The example user journeys described in Table 1 illustrate the flexibility and benefits an e-SafetyCase offers.

Table 1. Example e-SafetyCase user journeys

No	User	Requirement/Benefit
1	Service Technician	Familiarize himself/herself on the important factors in executing a safe transfer from marine vessel to offshore wind turbine whilst transiting out to the wind farm using the e-SafetyCase downloaded onto a mobile device.
2	Marine Controller	Refresh himself/herself of all the SCAs they are responsible for to ensure he/she performs these roles effectively.
3	Maintenance Technician	Find the method statement to allow him/her to complete one of their designated SCAs and watch an instructional video on how to complete that task.
4	HSE Manager	Assess potential vulnerability to a recent incident on a similar wind farm by interrogating the resilience of the barriers included within the relevant Bowtie diagram based on the findings of the incident investigation.
5	Regulator	Understand how the risk of fire is being managed within a wind turbine to form a view as to whether this risk is being managed ALARP.

What it is not....

It is important to recognize that the e-SafetyCase described here does not present a live risk monitor or risk “dashboard”; something which, confusingly, is often termed an e-SafetyCase.

This is something industry clearly has an appetite for but is a very different challenge to that being explored here. Work is ongoing to develop tools to do this and hopefully these could ultimately be blended with this e-SafetyCase concept to provide a complete, holistic solution to managing risk within a major hazard facility.

4 What benefits does an e-SafetyCase offer?

It goes without saying that an e-SafetyCase has to do what a traditional paper based Safety Case does; demonstrate risk is reduced ALARP. The real benefits of an e-SafetyCase come from the extra things it can do. It's a Safety Case, but not as we know it!

As previously described, this e-SafetyCase provides multi-dimensional links to allow any individual involved or interested in the safe operation of the offshore wind farm to quickly and easily access the information they need to perform their role effectively. It houses all of the information to ensure, demonstrate and support safe, reliable operation in one place and puts this at the fingertips of the right people in a format that suits them. This delivers three key benefits:

- It brings the information within the Safety Case to life in a simple, interesting and engaging way promoting a better understanding of safety and better safety compliance across the facility. As opposed to a traditional paper-based Safety Case, which tends to sit on a shelf gathering dust, an e-SafetyCase can actually promote and drive safe operations by encouraging feedback from the field that allows it to be improved and optimised to the benefit of the overall facility.

- By providing access to all of this information in a very simple and intuitive way, it becomes helpful and useful to anyone involved in the operation of the facility, not just those concerned about safety. When information is useful, it is well used, and so there is a natural incentive to keep it up to date. As a result, the e-SafetyCase evolves with the facility, ensuring there is no disconnect between the Safety Case and operations; they become one and the same (see Figure 8, with apologies to London tube stations). This overcomes the huge limitation of traditional paper-based Safety Cases; they are a burden to maintain and if they do not evolve with the facility, safety is inevitably compromised.

Figure 8: An e-SafetyCase bridges the gap between operations and safety

- By linking directly to traditional management system documentation such as procedures, method statements, job descriptions, etc., the e-SafetyCase provides a seamless link from the arguments as to why something should be safe right through to the documentation provided and responsibilities discharged to make sure it is actually safe. This is a significant improvement over traditional paper-based Safety Cases which tend to stop with the "why?" – the snapshot at that moment in time when the case was written.

So, in summary, the key benefit of an e-SafetyCase is that it:

- provides an accessible, engaging and intuitive basis of safety which proactively drives and encourages safe operations;

- naturally evolves hand-in-hand with the facility as routine business; and

- provides a seamless link between hazards and the actions people need to take to ensure risk is reduced ALARP.

Effortless safety – is that not too good to be true?

5 So why aren't all Safety Cases electronic?

The e-SafetyCase approach has been around for some time, but has struggled to gain widespread traction across industry, potentially due to misconceptions about the complexity and hence cost of development or the need for bespoke software. In reality, it's just about presenting information in a different and more accessible way, created using simple software found on most computers. What's more, this format can be applied to any industry for any facility. The cost of generating an e-SafetyCase is no greater than a paper-based Safety Case. All the underlying analysis, studies, etc. remain identical, it is just the final presentation that is different. If the decision to present the Safety Case in an "electronic" format is taken early, then it can be developed at no cost or programme penalty compared with a traditional paper-based version.

Another possible reason for not developing an e-SafetyCase is the concern that maintaining and updating it is too difficult and time consuming. Although updating the e-SafetyCase may seem more onerous than a traditional Safety Case, with a good management of change log and guidance, updating the e-SafetyCase should be no more taxing than updating a traditional Safety Case. It may just need someone with a different skill-set. Indeed, as previously demonstrated, an effective e-SafetyCase will be so useful and effective it will naturally evolve hand-in-hand with the facility as a natural part of routine business.

There's a danger the e-SafetyCase grows into a catch-all document due to the extra functionality it offers, however the intended focus of the e-SafetyCase should not be lost; it should not replace existing management systems or document systems but rather act as a link and interface into these complementary existing systems.

6 Further development and future potential

6.1 e-SafetyCase App

The offshore wind farm e-SafetyCase we have described here was built in PowerPoint with interactive links and then converted to a PDF. Whilst very simple to develop and use and low-cost to deliver, primarily because it uses standard software everyone has access to, all of the logic built into the Bowties had to be recreated within PowerPoint. This is obviously time consuming, introduces the potential for errors and makes updates more difficult.

To provide a more effective, long-term solution, we have developed an App which we can use to build an e-SafetyCase for an asset, facility, etc. The App works on PC, tablet and mobile phone (PC/Android/iOS). The App "sucks-in" the Bowtie native files which cover the major accident hazards for that facility (e.g. Offshore Wind Farm, Oil & Gas Offshore Platform, Onshore Refinery Plant, etc.) and provides a simple, engaging and intuitive way of navigating around this information in the same way as we have demonstrated here for the offshore wind farm.

So, in the analogy below, the Bowtie files are the image carousel and the e-SafetyCase App is the viewer. This therefore provides all of the benefits of the PowerPoint e-SafetyCase that we have demonstrated here, but with huge time and cost savings as it is automatically generated by the Bowties, thereby avoiding errors and making future updates and changes effortless as they simply evolve out of the Bowties which are maintained through the life of the asset/facility.

Rather than being a "product" which we sell, we use this App as an internal tool to help create an e-Safety Case for our clients. All Safety Cases are ultimately bespoke and are tailored to present a "Case for Safety" for an individual asset, facility, etc. Our VIVA App provides this functionality, enabling a bespoke e-SafetyCases to be built for each client and which is delivered to them in App form to use and maintain as they wish.

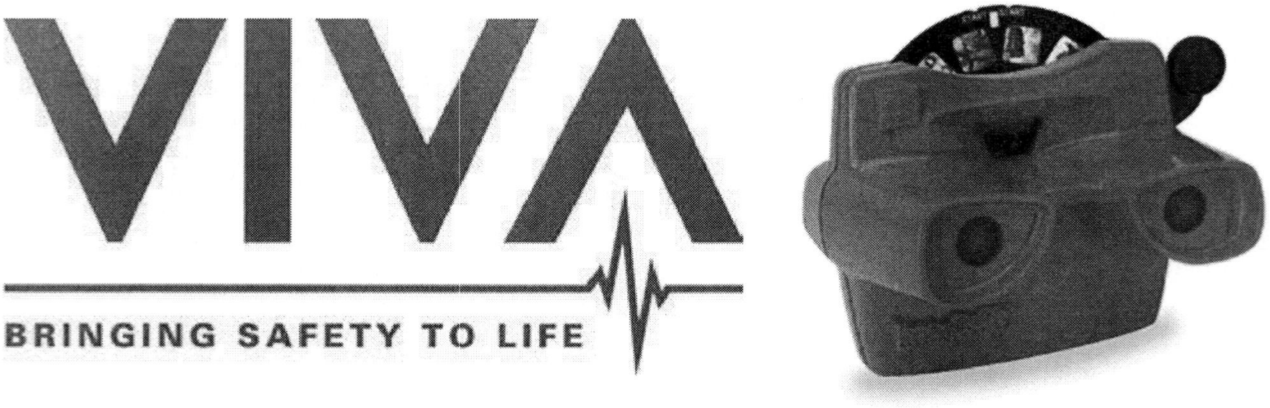

6.2 Blending with Digitalization

The e-SafetyCase concept has significant scope for further development and improvement by embracing wider technological innovations. A good example of this potential is the work we are doing to blend the e-SafetyCase with Virtual Reality (VR) to truly bring the Safety Case to life by immersing the user within the actual facility and overlaying Safety Case information on the plant itself, thus removing the disconnect between the facility and its Safety Case.

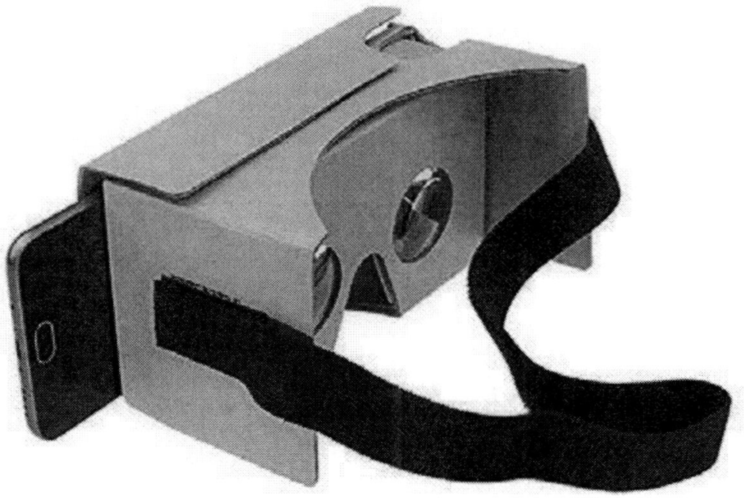

Figure 9. Google's Cardboard VR Goggles

VR technology has evolved significantly over recent years and is now available in very affordable and accessible formats such as Google's cardboard VR goggles (see Figure 9). These goggles incorporate special lenses and accommodate a standard mobile phone. The lenses translate moving images played on the phone's screen via compatible Apps into a 3D VR experience. The positional sensors within the phone are synchronized with the image being played on the device such that when the user tilts their head or turns around, they see the corresponding view creating a highly realistic, 360° immersive experience. In addition, a button on the side of the goggles taps the mobile phone screen allowing the user to select options built into the video footage played on the mobile device providing an interactive experience.

We have developed a prototype e-SafetyCase augmented with VR for a Wind Turbine Fire Safety Case. The Safety Case is viewed via an app downloaded onto a standard tablet device. Where VR enhances the Safety Case, a VR icon appears, which, when clicked, launches a VR snippet on a separate mobile device located within the Google goggles. The user can then put on the goggles and actually experience that element of the Safety Case in real, absolute terms. Some examples of the enhanced functionality this delivers include:

- A VR icon appears next to a SCE within a Bowtie diagram. Clicking the icon within the Bowtie launches a VR snippet which allows the user to see where that feature exists within the wind turbine in 3D terms and to understand its size, scale, configuration, etc. just as if they were physically there within the turbine itself.

- A VR icon appears next to the SCAs assigned to a responsible person. Clicking the icon launches a VR snippet which allows the user to see that feature so that they can experience in real terms how they must go about completing that activity.

- A VR icon appears next to the key stages of the various Escape, Evacuation and Rescue (EER) routes from the wind turbine in the event of all credible fire scenarios. Clicking these icons launches VR snippets which allow the user to experience the complete escape route, including;

    - the physical route they must follow;
    - the location of EER equipment (e.g. immersion suit, harness and descent device to allow an external escape from the nacelle of the turbine through the rear escape hatch as a last resort);
    - the time they are expected to take to progress through the route to ensure they can escape before it is compromised (based on findings of the EER study, supported by fire and smoke modelling);
    - the things they need to do at each stage (e.g. put on harness, connect to decent device, clip device onto anchor point, open escape hatch, etc.), and;
    - the extent to which visibility is likely to drop as they progress through the escape route by simulating the results from the supporting fire and smoke modelling.

  This delivers two key benefits:

    - It allows operators to familiarize themselves with EER arrangements – providing an excellent training tool.
    - It provides confidence (or otherwise) in the adequacy and acceptability of overall EER provision by allowing users to test, in a safe, virtual environment, whether escape can be completed safely before the overall route is compromised, thereby reinforcing the integrity of the case. Don't just read about why something is safe, experience it!

However, the direction of travel is not just one way. When wearing the goggles to view the VR snippets, the user can click the button on the side of the goggles to access related information which can be viewed back on the tablet. For example, SCEs are highlighted within the VR model and when clicked (via the button on the goggles) a simple menu opens-up allowing the user to select and then view (back on the tablet) either:

- The Bowties where that SCE is claimed;
- the SCAs associated with that SCE; or
- instructional videos associated with that SCE.

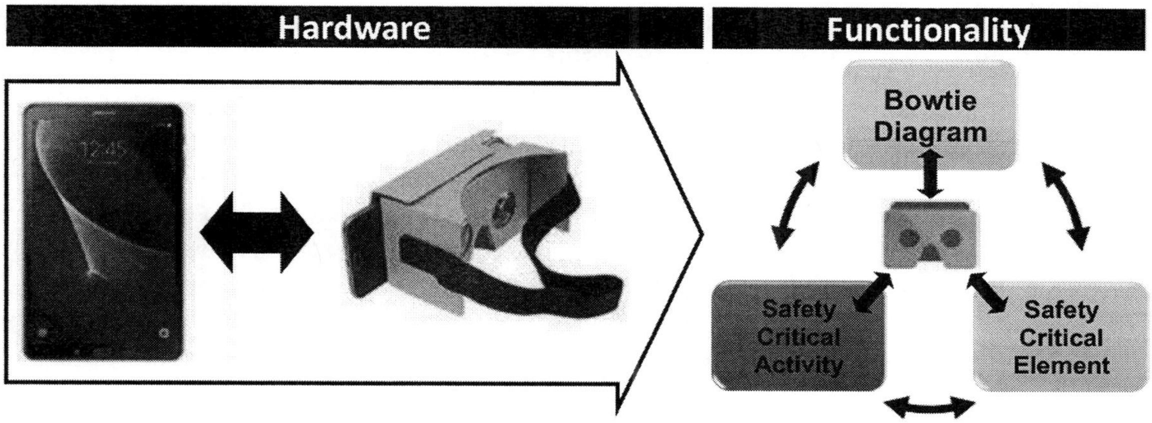

Figure 10. e-SafetyCase augmented with Virtual Reality

As illustrated in Figure 10, this two-way connectivity greatly improves the e-SafetyCase user experience from the version previously demonstrated (see Figure 7). The VR complements the e-SafetyCase, further bringing it to life, whilst also being a real life navigational tool for the Safety Case itself making it even more accessible, intuitive and easy to navigate.

The Holy Grail for any Safety Case is for the plant itself to show why it is safe. An e-SafetyCase augmented with VR goes a long way towards achieving this.

7 Perceived limitations of e-SafetyCase versus paper-based Safety Case

As the saying goes; "the proof of the pudding is in the eating!". As with any other innovation, e-SafetyCases will only become the established norm if they are better than what already exists. They must never be championed as an academic or theoretical crusade - they will rise or fall based on this very practical test. So, before significant time, effort and money is spent on the development of e-SafetyCases, as any good Safety Case author should always do, we must critically challenge whether they are indeed better than traditional paper-based Safety Cases by considering the following potential limitations:

1. **Ease of accessibility – can they really be "accessed by anyone, anywhere, anytime"?**

   By making the e-SafetyCase available as an App which can be downloaded onto a standard mobile device, yes it can. Most people have an electronic device which they could download the App onto. Also, the price of this technology is dropping all the time and so the cost to the company of providing all Safety Case users at a facility with an intrinsically safe tablet would be small in comparison to the cost of developing the Safety Case itself.

2. **An e-SafetyCase provides lots of peripheral benefits, but does it actually deliver on its core function – demonstrating a safe position?**

   An e-SafetyCase is founded on the same risk management process, studies, analysis, assessments, etc. as a traditional paper-based Safety Case. The only difference is how this information is presented, with these differences delivering the wider, peripheral benefits described elsewhere within this paper.

3. **A regulatory Safety Case is a strictly controlled document and for very good reason. How can an e-SafetyCase deliver the same level of control?**

   One thing a traditional paper-based Safety Case and an e-SafetyCase have in common is they are both deliverables that have a defined status and identity. As such, change management can be controlled in an identical way using existing processes. These will ensure revisions are developed, implemented, reviewed and approved by competent people, with updates strictly controlled to ensure that people only use the latest, approved version. Whether the Safety Case is presented on paper or in electronic format is immaterial in this regard. Indeed, the days of a SafetyCase being a printed book issued to a small number of approved users are (largely!) long gone. The traditional paper-based Safety Case has morphed into an electronic document which can be downloaded off servers and e-mailed around organizations. An e-SafetyCase presents no greater challenge to that which is already being managed.

   Making the e-SafetyCase available as an App does introduce additional challenges, particularly around security, but other businesses, with similar sensitive information which is version critical, manage these challenges effectively, e.g. finance/banking and e-commerce. In these examples, these challenges are managed in the public domain (the App is publicly available). There is no need for an e-SafetyCase to be publicly available with access only granted to approved users, thereby introducing another level of control.

4. **Regulated industries expect a traditional paper-based Safety Case. This is what they are used to and, indeed, the regulator has invariably shaped the format, structure, content, etc. of these. Will regulators accept an e-SafetyCase?**

   An e-SafetyCase does exactly the same job as a traditional paper-based Safety Case; it provides a robust demonstration of safety. The difference is in how this information is presented. With an e-SafetyCase being more accessible, intuitive and easy to navigate than a traditional paper-based Safety Case, it should make a regulator's job easier and so should be welcomed as a positive development. Indeed, it is a fundamental expectation of any regulator that operators embrace ways of reducing risk. With an e-SafetyCase providing an accessible, engaging and intuitive basis of safety which proactively drives and encourages safe operations, it should be viewed favorably, provided the challenges are adequately managed. Further support, engagement and buy-in would be gained by actively involving the regulator in the development of the e-SafetyCase concept, allowing it to be developed to meet their specific needs.

   Regardless of the above position, it is worth remembering that an e-SafetyCase is founded on the same risk management process, studies, analysis, assessments etc. as a traditional paper-based Safety Case. It would therefore be simple to also present it in the established, traditional format accepted within that industry. An e-SafetyCase which can be viewed in "traditional" and "electronic" modes would therefore provide the best of both worlds.

5. **Software projects are notorious for mushrooming in scope and complexity and blowing budgets and programmes. Would the development of e-SafetyCases suffer these problems?**

   An e-SafetyCase template would be developed as a tightly controlled project based around a detailed specification developed by the Safety Case users. This template would then be used to generate all future Safety Cases ensuring a consistent, proportionate approach. Clearly the scope and complexity of the e-SafetyCase could mushroom, but this can be readily managed through this one-off project. From that point onwards, the established blueprint will prevent individual e-SafetyCases being overly complex. This is actually an improvement over traditional paper-based Safety Cases where experience shows that they can tend to grow and expand unnecessarily from project to project, and also over the years as they often pass from one owner to the next and more information is bolted on in an ad-hoc way.

It will ultimately be the users (the designers, operators, managers, regulators etc. of major hazard facilities) who will decide whether the e-SafetyCase has a future.

There is clearly evidence that an e-SafetyCase can be at least as good as a paper-based Safety Case in terms of its core purpose of demonstrating a safe position. Adding in the wider, peripheral benefits it offers, there would appear to be a strong case to offer these users the chance to trial the concept so that they can make the ultimate decision. Let the people decide!

8 Conclusions

In today's world where answers to most questions can be Googled instantaneously, isn't it time the safety world caught up? The development of an e-SafetyCase is no more onerous than a conventional Safety Case, it's just different. Maintenance is no more onerous either. Indeed, an effective e-SafetyCase will be so useful and effective it will naturally evolve hand-in-hand with the facility as routine business.

Climbing the learning curve is a small price to pay for the considerable benefits this concept offers; providing an accessible, engaging and intuitive basis of safety which proactively drives and encourages safe operations.

Taking things further, the e-SafetyCase can be blended with technological innovations such as Virtual Reality to further bring it to life. This kind of ambitious approach has the potential to change the landscape of safety management in high hazard industries.

9 References

[1] The Nimrod Review
An independent review into the broader issues surrounding the loss of the RAF Nimrod MR2 Aircraft XV230 in Afghanistan in 2006
Charles Haddon-Cave QC
https://assets.publishing.service.gov.uk/government/uploads/system/uploads/attachment_data/file/229037/1025.pdf

[2] Buncefield: Why did it happen?
The underlying causes of the explosion and fire at the Buncefield oil storage depot, Hemel Hempstead, Hertfordshire on 11 December 2005
http://www.hse.gov.uk/comah/buncefield/buncefield-report.pdf

Procedure Simulators That Ensure Your Procedures Are Accurate and Operators Are Fully Competent

Presenter/Principal Author: Francis Montemurro
Affiliation: Voovio Technologies
Address: 1888 Stebbins Dr, Ste 100, Houston TX 77043
Email address: fmontemurro@voovio.com

Other Presenter/Author: Christian McDermott
Affiliation(s): Voovio Technologies
Address: 1888 Stebbins Dr, Ste 100, Houston TX 77043
Email address: cmcdermott@voovio.com

Prepared for Presentation at
American Institute of Chemical Engineers
2020 Spring Meeting and 17th Global Congress on Process Safety
Virtual
April 18 - 22, 2021

Copyright © 2021 Voovio Technologies US, Inc. All rights reserved.

Procedure Simulators That Ensure Your Procedures Are Accurate and Operators Are Fully Competent

Presenter/Principal Author: Francis Montemurro
Affiliation: Voovio Technologies

Other Presenter/Author: Christian McDermott
Affiliation(s): Voovio Technologies

Prepared for Presentation at
American Institute of Chemical Engineers
2020 Spring Meeting and 17th Global Congress on Process Safety
Virtual
April 18 - 22, 2021

Keywords: Field Operations, Human Error, Procedure Clarity, Procedure Execution, Procedure Simulator, Execution Errors, Unplanned Downtime, Training, Simulation, Operational Discipline, Competency.

Abstract

It is well-documented that many manufacturing unplanned events could have been prevented if procedures were executed properly. Ensuing investigations cite 'failure to follow procedure' as the event root cause resulting in next step recommendations, including additional operator training and discipline. However, based on data we collected from across North America, we consider "Procedure Clarity" a major source of error that is overlooked by the industry.

We believe there are three things that need to happen in order to drive real procedure clarity for operators of all levels:

1. You need clear and concise procedures;
2. You need the ability to visualize the procedure action steps;
3. You need the ability to practice and learn the procedure in a standardized way (learn by doing).

1 Procedure Clarity versus Operator Execution

ARC Advisory estimates that downtime costs the process industries somewhere in the area of a trillion dollars a year.[1] At Voovio, we believe the cost of inadequate procedure clarity and field operation execution errors is substantial, based on our own experience of working with customers across the petrochemical, specialty chemical and refining industries. An ASM Consortium study put the losses associated with unplanned downtime in the US Petrochemical

Industry at $10-20 billion annually.[2] Indeed, another ASM Consortium study[3] on human error and root cause of incidents identified causes of abnormal events as:

- procedure not followed : 51%
- procedure wrong: 40%
- procedure followed incorrectly: 6%
- Other: 3%

It is well-documented that many manufacturing unplanned events could have been prevented if procedures were executed properly.[4] Ensuing investigations cite 'failure to follow procedure' as the event root cause resulting in next step recommendations, including additional operator training and discipline. However, based on data we collected from across North America, a more systemic failure could be at the root of these unplanned events.[5] When an operational execution error occurs during a process or equipment startup, shutdown, changeover, process loss out of primary containment or other event, how often is "Procedure Clarity" considered the root cause? We consider it as a major source of error that is overlooked by the industry.

Operating procedures originate during capital project installation and are developed using the technical process design basis. Procedures include safety warnings, specific instructions, and details required to complete the desired outcome of the titled procedure. Following a successful project startup, the initial operating procedures are redlined, updated based on startup and practical experience, and institutionalized as part of the site's management of change (MOC) process.

We have identified two major reasons for suboptimal procedure clarity:

1.2 Procedure does not reflect operator practice

We analyzed 300 procedures from Customers across the US and Canada (between 2018-2020), broken down into the categories below in Figure 1. Interviews with customers and our procedure analysis shows a significant number of procedures do not match operator practice, a foundational element for Operational Discipline. 90% of the procedures analyzed contained generalizations.

Figure 1 – distribution of ~300 procedures analyzed by Voovio[6]

Type of Procedure	%
Abnormal	1.40%
Emergency Operations	2.46%
Emergency Shutdown	0.35%
Equipment Swaps	0.35%
Initial Startup	1.05%
Loading/Unloading	4.91%
Maintenance Overhaul	1.05%
Normal Operations	3.16%
Normal Shutdown	1.75%

Shutdown	27.37%
Startup	38.60%
Startup and Emergency Shutdown	3.16%
Startup and Normal Operations	2.81%
Startup and Shutdown	5.61%
Startup following a Shutdown	4.91%
Temporary Operations	1.05%
TOTAL	**100%**

When surveying manufacturing process subject matter experts (SMEs), the written procedures vary from the unit's established practice with some practices varying from shift to shift either due to training inconsistencies or a lack of standardized execution. This finding was previously documented as the cause for 23% of manufacturing's downtime losses by the Ron Moore Group, in Figure 2 below.

Figure 2 – Categories of downtime losses in process industry.[7]

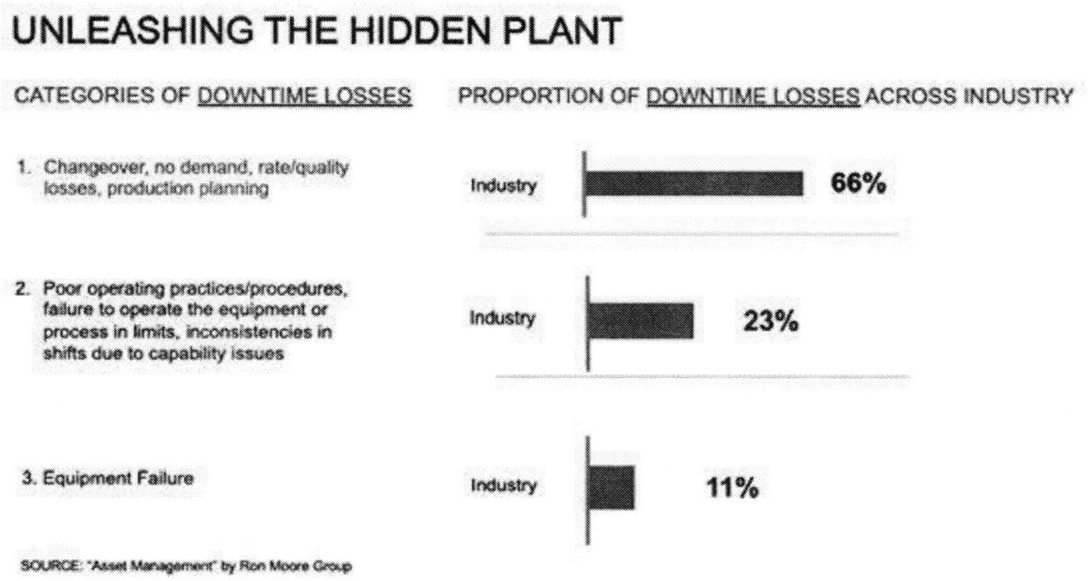

1.3 Ambiguities and generalizations

Using procedures, manufacturing operators train on how to operate the process, unit or plant safely and effectively. Operators are challenged to learn and qualify on hundreds and hundreds of pages across a range of procedures. Often, these procedures are written with a lot of technical content, generalizations and ambiguities that assume clear understanding by the operator. Procedures are typically written by engineers or experienced operations personnel (SMEs), individuals very familiar with the process, who tend to overlook the basic details of execution believing the details are largely understood. Below are some examples from actual manufacturing procedures (we have changed the equipment tag numbers to ensure anonymity).

Example: Procedure line with additional peripheral information and options on how to execute the procedure;

"Level indicator transmitters (ABC-111 and DEF-222 at A-C-9 and GHI-871 and GHI-872 at A-C-8) provide an initial inventory in these vessels. To provide initial inventory in A-C-7 and A-C-6, 30-psig condensate control valve (FV-4447) can be manually opened or put in Automatic at 30%. The steam regulator for A-C-6 is located beside the CO2 plant." [8]

The first and third sentences are not needed for procedure execution; thus, may add uncertainty to the operator. The second sentence gives the operator options; leaving to question what the preferred best method is to inventory the system.

Examples: General, ambiguous information in procedures;[9]

- *"Close all bleeders and open-ended lines"*
- *"Blow product return lines and control valve loop with blend gas from A-123 take-off manifold to ensure lines are clear then isolate from all equipment"*
- *"Outside operator will verify manual valves on feed system are lined up"*
- *"Ensure cooling water is lined up to all coolers and exchangers that require cooling water"*
- *"Manually close all condensate controllers outside"*

An operator is expected to follow the above procedural statements and remember where 'all' the components are located while executing the one procedure line item. In the 1990s and early 2000s, this problem was masked by workforce experience, shift teams had many individuals with 20 to 30+ years of operating experience, thus procedure clarity was not important.[10] With the gap in experience, the lack of procedure clarity, and few operator aids to ensure execution accuracy, incidents and unplanned events will continue to occur.

2 Impact of procedures during operator training

Typically, once the operator reads the required procedures, shadows an experienced operator for the designated time and completes some form of assessment, he or she is 'signed-off as qualified' in the job. However, industry data indicates the operator has only actually executed or practiced around 40% of the procedures due to the fact that equipment is always running and plant operation priorities.[11] Post qualification, the operator is released to work the job. Thus, procedure quality and, more importantly, specificity is the building block for developing operator competency.

Procedure formats and templates vary from company to company and even from site to site. Some companies have implemented company or site-wide digital solutions, involving many technical writers, significant company involvement and subsequent procedure conversions. Other companies maintain procedures manually with internal document control systems. Depending on the implemented solution, sites require operators to print procedures prior to execution and sign or check off steps as they are completed, or with digital formats, operators carry tablets in the field to use as a reference tool to read the procedure prior to executing each line item. Both formats require operators to read procedures for understanding. It is well documented in learning and development forums that individuals learn and retain ~90% using simulation and doing versus ~10% when reading procedures; see below. The industry challenge is to move operator training, proactive development, down the pyramid to increase individual retention and reduce an operator's time to full competency.

Figure 3 – the Learning Pyramid.[12]

The Learning Pyramid or Cone of Learning

Ultimately, to see a step-change improvement in operational discipline, procedures should be developed, implemented, and maintained based on successful execution by the end-user… the operator; and they should be clear, concise and action-based.

We believe there are two best practices during project development to ensure accurate procedures and standard operator execution. First, a Forward Engineer (creator of the procedure simulator)[13] performs a Line Walk for selected customer procedures with the resident SME to document procedure execution accurately using photographs and creating an action-based checklist. This visual walkthrough identifies gaps in the written procedure and the actual operating practice. All deviations from the documented procedure are reviewed and approved using the site MOC process. One extreme example, but not uncommon, is the following **<u>one</u>** procedure line step:

Pre-Startup Checklist SOP: "1. *Using 70-psig nitrogen, purge all equipment and process lines to remove oxygen and to ensure there are no leaks or bleeders left open before introducing feed to the unit.*"

Once the procedure had been simulated (following the Line Walk in the field with the SME and Forward Engineers) this single step consisted of 112 actions that the outside operator needs to perform across a sizeable section of the unit including:

- Interacting with bleeder valves 46 times
- Interacting with inlet/outlet valves 26 times
- Interacting with supply valves 8 times
- Performing line purges and other tasks

Operators confirmed that there is no 'agreed' way to perform this step, and there are variations in the way this step is performed from shift to shift. Typically, procedures as ambiguous and generic as in this example would be updated to better reflect the reality of what happens in the field, therefore providing a 'standard' for the unit. An action checklist detailing these specific actions is printed and submitted to the Operations Manager and Procedure Owner for review, with a red-lined version of the original procedure. The site's MOC process is followed.

Figure 4.

Action Checklist detailing every action the operator needs to perform

Customer Procedure Document

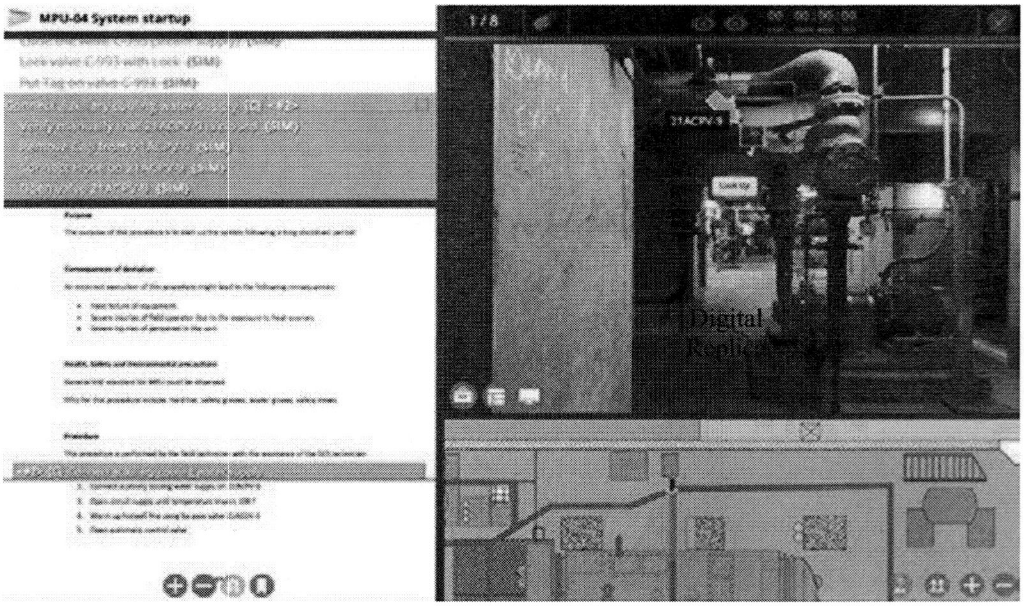

Additionally, using Voovio's patented technology, users can perform the newly documented procedure in an immersive photo-based environment (called the 'Digital Replica') and interact with and manipulate (open, close, push, etc.) on-screen components. This method of simulating operation allows operators, engineers, production specialists and others, the chance to **visually**

verify procedure accuracy. A simple and extremely effective way of validating and certifying the written procedure.

The second best practice is the addition of action steps, better known as a checklist. This well-documented methodology has significantly improved safety performance in air travel, medical procedures and other work environments.[14] During the Line Walk with the SME, the Forward Engineers confirm each action the operator is expected to complete. We regularly discover there are multiple actions within a single procedure line. Below are two examples to highlight this.

- The procedure line item, *"Close all bleeders and open-ended lines"*[15] resulted in 12 action steps because there were 12 bleeders and open-ended valves to be closed to ensure no contents left primary containment and/or caused an environmental incident.
- The procedure line item, "*Manually close all condensate controllers outside*"[16] resulted in 20 action steps to close 16 valves.

The process of simulating a procedure, with the Line Walk performed by an SME and Forward Engineer, results in (on average) 2.4 times more action steps than the number of procedure line items. Procedures are complemented by an action-oriented checklist, thereby making them clearer and improving the human reliability component when executing them in the field. A summary of past projects is depicted in the graph below (Figure 5).

Figure 5 – Procedure steps before and after a line walk for simulation.[17]

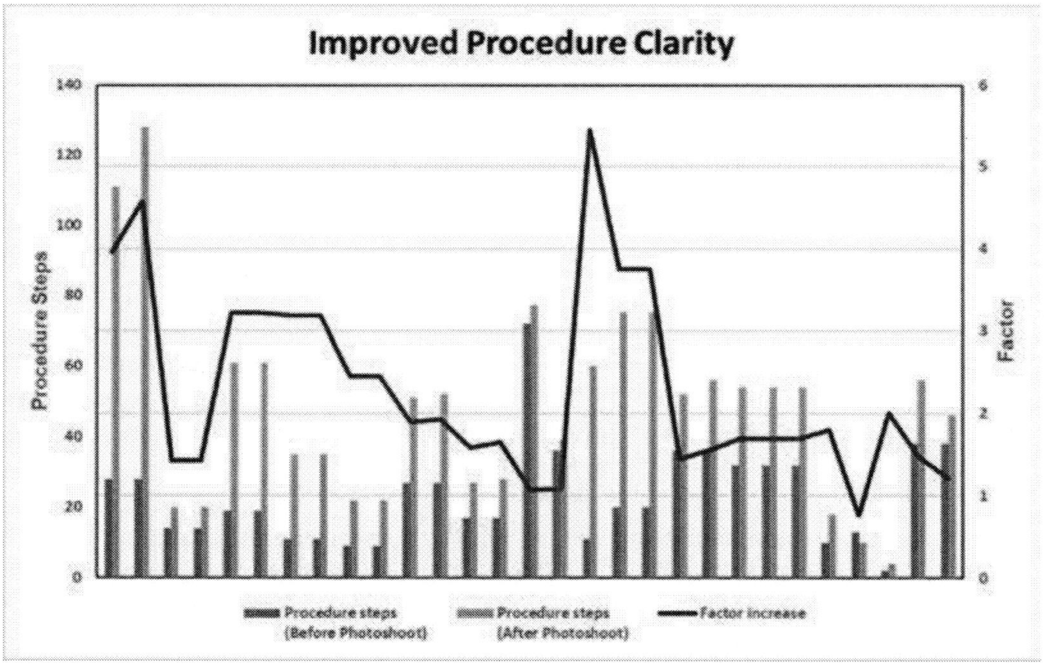

Voovio's interactive and comprehensive Forward Engineer (FE) Line Walk process eliminates procedure generalities and ambiguities resulting in clear and concise action steps. On average, procedure steps increase by a factor of 2.4

3 Digital Replica: environment required to 'simulate' and 'verify' procedures

Simulation requires a visual representation of the assets, which is displayed on a screen. Simulation technologies available include CAD-based technologies and Enhanced Reality technology. This paper considers simulation and procedure line walks that use Enhanced Reality technology, which was specifically designed for the manufacturing environment.

The Digital Replica is the basis of the simulated environment. It is a collection of standard photographs that are arranged in such a way that the user feels as though they are walking seamlessly throughout the process environment and able to interact with the equipment. In the Procedure Simulators, written procedures and the actions in the checklist are linked to the Digital Replica. Users are able to review and perform each action in the Digital Replica as if they were in the real plant. The Procedure Simulator Test Module monitors performance as users walk around and interact with the equipment, providing immediate feedback on incorrect and correct actions.

Operations, engineering, maintenance and other critical personnel can utilize the Digital Replica and Simulators to verify if the procedure is correct. It is not uncommon that a procedure is updated and improved after SMEs have utilized the simulation. Another simple example of procedure improvements thanks to using the Digital Replica and Simulators is an inspection procedure for a chlorine railcar. In order to perform the checks, operators would walk around the railcar almost 20 times with multiple inspection points. Following the simulation, which highlighted this unnecessary inefficiency to the senior trainers and SMEs, the procedure was updated to produce a more efficient and effective route for inspections.

4 How is this approach economically viable?

The criteria for selecting procedures to simulate include the frequency of incidents, including safety, environmental or product quality losses which can be traced back to operator error. In addition, procedures that are not frequently executed are also good candidates, because operators may not have the opportunity to experience or train on them unless they happen.

The Line Walk process with a Forward Engineer ensures that a procedure is correct by walking step-by-step with the Subject Matter Expert. In surveys conducted with end users (operators) between 2018-2020, 100% of respondents believed the Line Walk process and Simulated Procedures makes it easier to learn and understand procedures than their current (traditional) training methods, provided clarity on aspects of their job that they did not have before, and will increase compliance with operating procedures in their area.

In Figures 6-8, data taken from Voovio customers show the financial impact of downtime losses associated with specific equipment and procedures. This is a very small sample from different operating units in both Refining and Chemical plants. A single refinery or plant may have thousands of similar procedures. These incidents happen frequently across the industry. Having accurate procedures is foundational to having good operational culture and discipline. While clearer procedures will help reduce incidents, organizations need to provide operators with the tools and training to be successful.

In each example below, the cost of simulating the procedures (including line walks) was far less than the cost savings realized, resulting in payback in less than one year.

Figure 6[18] - Data from a support process for refinery, providing technical gases. The site suffered from high attrition and unplanned downtime related to boiler feedwater (BFW) pump failures related to mis-operation. Thanks to improved procedure clarity and the ability for operators to practice and learn prior to execution, it increased organizational productivity and reduction of unplanned failures with an annual benefit of $ 88,000 thanks to simulating one procedure.

ASSET AVAILABILITY	TECHNICAL GASES
Annual Downtime Costs, $	220,000
Plant Trips, #	1
Cost per Plant Trip, $	100,000
Critical Equipment Failures, #	1
Critical Equipment Failure Costs, $	150,000
labor costs per event, $	15,000
contractor costs per event, $	-
Other Costs, $	5,000
Estimated Downtime Reduction, %	40
Net Program Yearly Benefit, $	88,000

Figure 7[19] – Data from a Petrochemical plant (Nonwovens). The unit suffered from incorrect process start-ups, low product yields due to operator execution errors. Several startup procedures were simulated, leading to improved procedure clarity and knowledge for the operators resulting in reduction of product losses, saving over $ 500,000 annually.

ASSET AVAILABILITY	PETROCHEMICAL
Annual Downtime Costs, $	1,142,000
Unplanned Events/year, #	1
Cost per Plant Event, $	1,116,000
Critical Equipment Failures, #	-
Critical Equipment Failure Costs, $	-
labor costs per event, $	15,000
contractor costs per event, $	10,000
Other Costs, $	1,000
Estimated Downtime Reduction, %	50
Net Program Yearly Benefit, $	571,000

Figure 8[20] – Data from a refinery, truck loading operations. The refinery aimed to reduce overtime to train contract drivers and reduce unplanned downtime on loading racks due to damage and spills/resulting in costly repairs and clean-up (~2.5 per month). 3rd party truck drivers are remotely trained, with clear actions, and assessed prior to loading leading to a reduction of incidents.

ASSET AVAILABILITY	TANK TRUCK HAZMAT
Annual Downtime Costs, $	2,675,700
Plant Events, #	27
Cost per Plant Event, $	99,100
Critical Equipment Failures, #	-
Critical Equipment Failure Costs, $	-
labor costs per event, $	-
contractor costs per event, $	-
Other Costs, $	-
Estimated Downtime Reduction, %	10
Net Program Yearly Benefit, $	267,570

In addition to the reduction of unplanned downtime due to poor procedure clarity and mis-operation, time savings are a key contributor to savings in operations training. Figure 9 reflects average speed to competency gains by using Digital Replicas and Procedure Simulators. Units see an average of 50% reduction in Subject Matter Expert (SME) time dedicated to training new operators and when onboarding operators, on average it takes 30 to 50% less time to be fully qualified.[21]

Figure 9 – Traditional Methods of Training versus Digital Replicas and Procedure Simulators

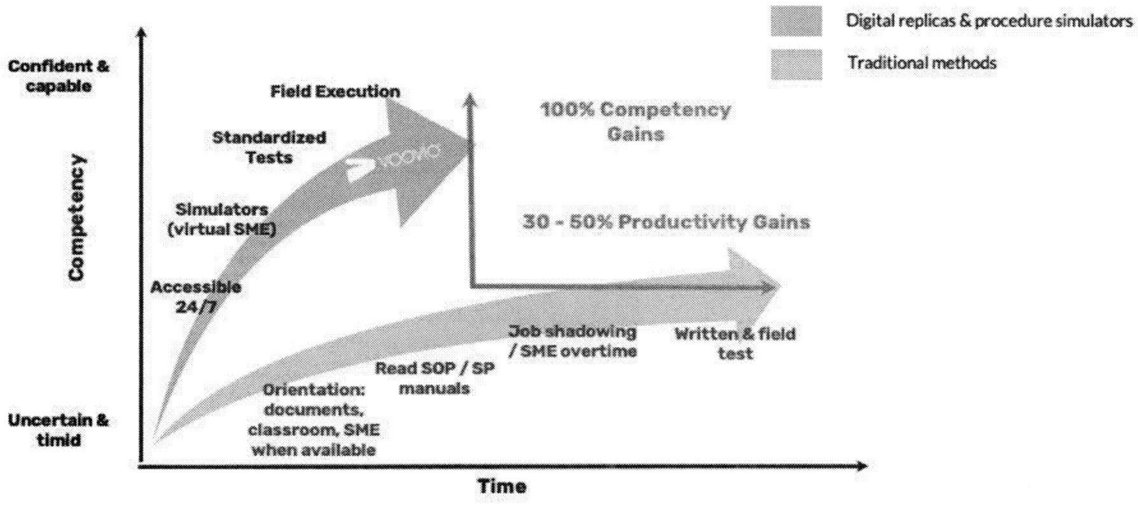

5 Conclusion

ARC Advisory concludes that "*the human role is paramount in reducing unplanned downtime... To avoid an incident, operators must receive better information and training to make the right decisions.*"[22] Whether this is for control room operators or field operators, the approach must be the same and the investment required to achieve this makes economic sense, based on the immediate payback achieved by preventing failures.

The data outlined in this paper, collected from Customers in North America, cite procedure clarity as a major source of error. The data shows there is a strong case for improving procedure clarity, so that the industry can improve how the 'knowledge' within those procedures can be transferred more effectively to operators themselves, thereby creating a standardized system for training and procedure execution, leading to a reduction in unplanned downtime.

Clear and concise procedures, standardized on plant operation that have objective assessments, are foundational building blocks to ensure field operator competency and deliver predictable process performance. Furthermore, operational discipline and speed to operator competency are and will remain competitive advantages in the Petrochemical and Refining industry due to the current and projected attrition slated for plant sites.

The road to execution excellence is complex, but by having the three foundational pieces we set out in this paper, manufacturing plant operators will have the tools to build their competency and deliver business results safely and successfully:

- Clear and concise procedures
- Ability to visualize procedure steps in the field
- Ability to practice standardized execution (*learn by doing*)

6 References

[1] *Process Industry Downtime and Key Performance Metrics, ARC Advisory, Larry O'Brien, April 2017.*
[2] *Process Industry Downtime and Key Performance Metrics, ARC Advisory, Larry O'Brien, April 2017.*
[3] *Improving Procedural Operations, ASM Consortium, Chris Morse / Andrew Ogden-Smith, Honeywell, September 2014.*
[4] *Based on reports from the Chemical Safety Board (CSB) and Francis Montemurro's own experience with investigations.*
[5] *Data collected from Voovio Technologies Customers between 2018 and 2021.*
[6] *Data taken from Voovio Customers, 2018-2020.*
[7] *Ron Moore is the Managing Partner of The RM Group, Inc. in Knoxville, TN. He is the author of Making Common Sense Common Practice: Models for Manufacturing Excellence (now in its 4th edition); and of What Tool? When? A Management Guide for Selecting the Right Improvement Tools (now in its 2nd edition), both from MRO-Zone.com; as well as Business Fables and Foibles and Our Transplant Journey: A Caregiver's Story; and over 50 journal articles.*
[8] *Extracted from a Standard Operating Procedure from a Voovio Technologies Customer, 2019.*
[9] *Extracted from a selection of Standard Operating Procedures from Voovio Technologies Customers, 2018-2020.*
[10] *Based on evidence collected from Francis Montemurro's 30+ year career with DuPont, in roles of plant management and operations leadership, including industry peers.*
[11] *Based on feedback from Voovio Customers, 2018-2021.*
[12] *Edgar Dale, The Learning Pyramid, https://en.wikipedia.org/wiki/Learning_pyramid, accessed on January 27th 2021.*
[13] *A Forward Engineer is an engineer by background, with experience in manufacturing who walks through procedures with an SME and captures the photographs required to produce the Procedure Simulators and documents the actions required to perform each step of the procedure.*
[14] *Data includes "The Checklist Manifesto – How to Get Things Right", Atul Gawande, 2010.*
[15] *Extracted from a Standard Operating Procedure from a Voovio Technologies Customer, 2018.*
[16] *Extracted from a Standard Operating Procedure from a Voovio Technologies Customer, 2019.*
[17] *Data collected from Voovio Technologies Customers between 2018 and 2021.*
[18] *Data taken from Voovio Customers, 2018-2020.*
[19] *Data taken from Voovio Customers, 2018-2020.*
[20] *Data taken from Voovio Customers, 2018-2020.*
[21] *Data extracted from Voovio Customers, 2018-2020.*
[22] *Process Industry Downtime and Key Performance Metrics, ARC Advisory, Larry O'Brien, April 2017.*

SPRING21
+17TH GCPS
A Joint AIChE and CCPS Meeting

¿Como alcanzar una protección contra incendio eficiente? El Fire Hazard Analysis (FHA) como herramienta de optimización de seguridad contra incendios

Alejandro Jimenez
Engineering Systems Inc.
8491 NW 17th Street, Suite 102
Doral, FL 33126
aljimenez@engsys.com

Prepared for Presentation at
American Institute of Chemical Engineers
2020 Spring Meeting and 17th Global Congress on Process Safety
Virtual
April 18 - 22, 2021

¿Como alcanzar una protección contra incendio eficiente? El Fire Hazard Analysis (FHA) como herramienta de optimización de seguridad contra incendios

Alejandro Jimenez
Engineering Systems Inc.
8491 NW 17th Street, Suite 102
Doral, FL 33126
aljimenez@engsys.com

Keywords: Fuego, Incendios, Protección, Mitigación

Abstract

La industria de procesos como petrolera, gas, química y farmacéutica maneja una cantidad significante de líquidos y gases inflamables capaz de producir incendios de gran tamaño. Estos incendios poseen el potencial de afectar tanto la seguridad del sitio de trabajo como la capacidad que posee la organización de continuar con su operación, afectando económicamente a la organización.

La protección contra incendio en las plantas de procesos se utiliza con el fin de prevenir y controlar los posibles daños ocasionados por dichos incendios. La protección contra incendios en estos tipos de plantas puede ser compleja ya que involucra diversos aspectos de operación, diseño y respuesta ante emergencias. De tal forma que se requiere de un estudio especializado que incluya todas las variables necesarias y permita entender directamente el riesgo de incendios con el fin de poder definir la mejor filosofía de protección. El Fire Hazard Analysis (FHA) es un proceso de identificación y análisis de los diversos eventos de incendios potenciales con el fin de consolidar todas las variables involucradas en el escenario, identificando la mejor estrategia de control y prevención de estos.

El FHA es una metodología basada en riesgo que incluye una etapa de análisis de riesgo, un taller de discusión, y finalmente la optimización de la filosofía de protección. El proceso se puede dividir en las siguientes etapas:

- Definición de objetivo.
- Análisis de riesgo.
    - Identificación de peligros.
    - Análisis de consecuencias de los eventos importantes.
    - Estimación de riesgo.
- Evaluación de desempeño de sistemas de protección.

- Taller FHA.
- Sistemas de protección y filosofía de protección.

Estas etapas permiten una progresión sistemática desde la identificación de peligros de incendios hasta la evaluación de la filosofía de protección encontrando el nivel óptimo de protección requerido para los riesgos presentes.

1 Introducción

La industria de procesos como petrolera, gas, química y farmacéutica maneja una cantidad significante de líquidos y gases inflamables capaces de producir incendios de gran tamaño. En el pasado estos incendios han destruido plantas de procesos impactando tanto al personal que trabaja en ellos, como la capacidad que tiene la organización de responder a sus clientes afectando su situación económica y comercial.

Debido al potencial que poseen estas instalaciones de generar un incendio, las mismas deben ser protegidas con sistemas de protección que controlen los efectos del incendio y disminuya los posibles daños ocasionados. Sin embargo, la protección contra incendios en estos tipos de plantas puede ser compleja ya que involucra diversos aspectos de operación, diseño y respuesta ante emergencias; diseñar e implementar sistemas sin considerar estas variables puede generar el fallo de controlar un incendio accidental. Un ejemplo claro de esto ocurrió el 19 de diciembre de 1982 cuando un incendio en la planta termoeléctrica de Tacoa en Venezuela tomó la vida de 160 personas, entre los cuales se encuentran 97 bomberos, diversos miembros de la prensa y vecinos de la zona. El sistema de protección contra incendio no incluía la protección de los tanques involucrados requiriendo que el personal de respuesta deba acercarse a los mismos. El evento evolucionó hasta la ocurrencia de un Boilover impactando a todas las personas en la cercanía. De tal forma que se requiere de un estudio especializado que incluya todas las variables necesarias y permita entender directamente el riesgo de incendios con el fin de poder definir la mejor filosofía de protección.

El FHA [1] (por sus siglas en inglés) es un proceso de identificación y análisis de los diversos eventos de incendios potenciales con el fin de consolidar todas las variables involucradas, identificando la mejor estrategia de control y prevención de estos. Este proceso permite la combinación de un análisis determinístico de modelación y simulación de incendios con el juicio y opinión de diversos expertos multidisciplinario para la definición de una filosofía adecuada de protección.

2 Análisis de Peligro de Incendios (FHA)

El FHA es una metodología basada en riesgo que incluye una etapa de análisis de riesgo, un taller de discusión, y finalmente la optimización de la filosofía de protección. El proceso se puede dividir en las siguientes etapas:

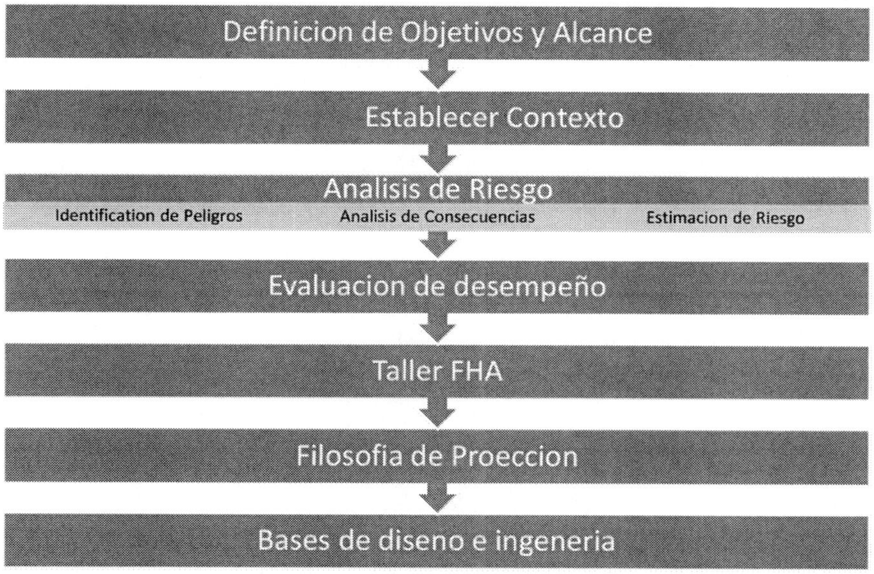

Figura 1. Proceso FHA

2.1 Definición de Objetivos y Alcance

Como cualquier estudio de ingeniería, el FHA debe definir un objetivo y alcance. De esta forma, los objetivos planteados pueden ser alcanzados utilizando eficientemente los recursos alocados para el estudio. El proceso es flexible y permite adaptar diversos objetivos, pero debe existir una definición clara del objetivo que se desea alcanzar y el alcance que abarca el estudio.

2.2 Contexto

El contexto incluye todos los parámetros, situaciones, y condiciones existentes relevantes a la instalación donde se va a realizar el estudio. La importancia de conocer el contexto radica en el entendimiento de como el entorno y la situación de la instalación interacciona con el riesgo. Por ejemplo, una instalación con operación remota va a limitar el potencial impacto al personal, pero la detección de incendios debe ser primordial para el reconocimiento del evento, en cambio una instalación con personal puede exponer al personal a los efectos del fuego, pero el incendio puede ser reconocido mucho más rápido.

2.3 Análisis de riesgo

El FHA empieza con una etapa de análisis de riesgo que permita entender el riesgo presente en la instalación. Este análisis de riesgo debe incluir una etapa de identificación de peligros, una evaluación de consecuencias de los incendios identificados y una estimación de riesgo.

2.3.1 <u>Identificación de Peligros</u>

Esta etapa es la primera actividad realizada en cualquier método basado en riesgo. La idea es identificar los peligros en la instalación y los posibles escenarios y eventos de incendios que podrían ocurrir.

Esta etapa es fundamental ya que, para poder proteger la instalación de posibles casos de incendios, se deben conocer los diversos eventos que puedan ocurrir y sus características. Por ejemplo, la ocurrencia de un incendio que no se tenía prevista o que se desconocía no podrá ser mitigado eficazmente ya que su ocurrencia tomará de sorpresa a los operadores y respondedores, los cuales no tendrán los equipos, recursos y medidas necesarias para el caso específico. Por otro lado, si el evento es conocido, se pueden implementar medidas que permita la acción adecuada y a la medida disminuyendo el daño que el evento pueda generar.

2.3.1.1 Características de la Identificación de Peligros

La identificación de peligros debe ser realizada para la gestión de riesgos de la instalación o proyecto, esta es realizada en un taller en el que participan diversas disciplinas, por ejemplo, ingeniería de procesos, ingeniería de instrumentación, operación, etc. Este taller debe ser liderado por un facilitador competente en identificación de peligros, independiente del personal encargado de la instalación o proyecto. Existen diversas metodologías estructuradas de identificación de peligros y es el deber del facilitador identificar la mejor metodología para el caso estudiado. Algunas metodologías usadas son Análisis preliminar de peligros, HAZID, What-If, HAZOP, FMEA.

Una de las metodologías más usadas durante el FHA es el HAZID, el cual se basa en identificar los peligros de la instalación estudiada en función a una lista de peligros guía usada en el estudio. Los peligros guías son posibles peligros presentes que deben ser evaluados. La desventaja de este método es la posibilidad de no identificar algún peligro por no estar listado, sin embargo listas estándares han sido desarrolladas y avaladas por diversas organizaciones, por ejemplo la guía de evaluación de peligros de CCPS[2] posee una lista de preguntas guías para diversas técnicas (p.ej. What-If) en su apéndice y el estándar ISO 17776[3] posee una guía de peligros en su apéndice (esta última, aunque fue desarrollada para instalaciones costa afuera (offshore), representa una buena guía que puede ser usada en otro tipo de instalaciones). Adicionalmente muchas organizaciones han definido listas a nivel corporativo que deben ser usadas en sus operaciones para la identificación de peligros.

Independientemente de la metodología utilizada, el procedimiento de identificación de peligros y riesgos consiste en dividir la instalación en áreas o secciones y evaluar si dichos peligros se encuentran en la sección discutida. Cuando se encuentre un peligro, se debe documentar el peligro, la causa o fuente del mismo y la consecuencia o efecto potencial que dicho peligro pueda producir. Por ejemplo, un incendio, explosión, etc. Adicionalmente, se estima el riesgo presente de forma cualitativa usando una matriz de riesgo y alguna mejora o recomendación identificada por el equipo.

Por otro lado, es importante que todos los peligros identificados y sus consecuencias sean descritos adecuadamente; debido a que estos peligros y consecuencias serán la base del estudio FHA. Los peligros deben estar descritos de forma clara y las consecuencias deben considerar el peor caso posible.

La Tabla 1 muestra un ejemplo de la hoja de trabajo usada en el taller para documentar los peligros y riesgos identificados. En esta tabla se muestra que las medidas de mitigación se muestran separadas de las medidas preventivas, esto se debe a que la protección contra incendio es una estrategia de mitigación y cualquier medida mitigativa listada servirá de soporte cuando se desarrolle la filosofía de protección contra incendios en etapas posteriores.

Tabla 1. Ejemplo de tabla de identificación de peligros

No.	Peligro	Causa/fuente	Consecuencia	Medidas Preventivas	Medidas de Mitigación	R	Recomendación

R – Riesgo definido por la matriz de riesgo utilizada

Ya que desde el punto de vista de FHA el interés es especifico en los casos de incendios, se debe realizar una revisión del resultado obtenido en el ejercicio de identificación de peligros elaborando una lista exhaustiva de todos los peligros y casos de incendios identificados. Dicha lista debe indicar sus características principales, incluyendo:

- Lugar de ocurrencia.
- Material involucrado.
- Posible tipo de incendio (charco, chorro, destello).
- Potencial de escalación del evento (p.ej., BLEVE).
- Potencial consecuencia a las personas, bienes y continuidad del negocio.
- Nivel de riesgo.
- Medidas de protección y prevención de incendios definidas.

Esta información debe ser resumida en un documento base. Este documento base tiene diversos nombres, pero suele llamarse **diagrama de áreas de incendios**. Este documento incluye la división de la instalación en áreas o zonas de incendios, y una tabla maestra con la información mencionada anteriormente por área de incendio; adicionalmente, suele indicar alguna otra información relevante, p.ej., clasificación, ventilación, ocupación, nivel de ruido, etc. La Figura 2 a continuación presenta un ejemplo de una tabla maestra del diagrama de área de seguridad, el cual es utilizado como base para el FHA.

AREA SAFETY CHART

Hazardous Area No.	Area Description	Hazards	Potential Major Accidents Events
Process Areas (PA)			
31 & 32	Cryogenic units	1) Flammable liquids and gas (NGL, Natural Gas, Propane, Methanol) 2) Combustible liquids (Lube Oil)	Jet Fire, Pool Fire, Explosion
41 & 42	Residue Compression Units	1) Flammable gas (Natural Gas) 2) Combustible liquids (Lube Oil)	Jet Fire, Pool Fire, Explosion
63	Backup Generator	1) Flammable gas (Natural Gas) 2) Combustible liquids (Lube Oil, Cooling fluid)	Jet Fire, Pool Fire, Flash Fire
21 & 22	Amine Sweetening Units	1) Flammable gas (Natural Gas) 2) Combustible liquids (Amine) - **operated above flash point**	Jet Fire, Pool Fire, Explosion
26	Amine Regeneration Unit	1) Flammable gas (Natural Gas) 2) Combustible liquids (Amine, Hot Oil) 3) Activated Carbon	Jet Fire, Pool Fire, Explosion, Dust Flash Fire/Explosion
91	Asset Gas Destruction (flare and thermal oxydizer)	1) Flammable gas (Natural Gas) 2) Asphyxian gas (Asset Gas - 91% CO2)	Jet Fire, Pool Fire, Explosion, Asphyxian media
14	Sulphur Removal	1) Flammable gas (Natural Gas) 3) Pure Spec (Filter Beds)	Jet Fire, Pool Fire, Explosion, Dust Flash Fire/Explosion
40	Outlet booster compression	1) Flammable gas (Natural Gas) 2) Combustible liquids (Lube Oil)	Jet Fire, Pool Fire, Explosion
50	TEG Dehydration	1) Flammable gas (Natural Gas) 2) Flammable Liquid (BTEX - Benzene, Toluene, Xylene) 2) Combustible liquids (TEG) - **Could be operated above Flash point**	Jet Fire, Pool Fire, Flash Fires, Explosion
80	Hot Oil Area	1) Flammable gas (Natural Gas) - blanket 2) Combustible liquids (Lube Oil)	Jet Fire, Pool Fire, Flash Fires, Explosion
70	Residue Metering	1) Flammable Gas (Residue Gas - 90% to 95% Methane)	Jet Fire, Pool Fire, Flash Fires, Explosion
10a	Slug Catcher	1) Flammable Liquids (Plant feed, Methanol)	Jet Fire, Pool Fire, Flash Fires
10b	Inlet Area	1) Flammable Liquids (Plant feed)	Jet Fire, Pool Fire, Flash Fires
11 & 12	Stabilization Units	1) Flammable liquids and gas (NGL, Natural Gas, Condensate) 2) Combustible liquids (Lube Oil, Hot Oil) 3) Activated Carbon (carbon bed filters)	Jet Fire, Pool Fire, Explosion, Dust Flash Fire/Explosion
71 & 92	Slug Tanks	1) Flammable Liquids (Condensate)	Jet Fire, Pool Fire, Flash Fires, Explosion
73	NGL Take away	1) Flammable Liquids (NGL)	Jet Fire, Pool Fire, Flash Fires, Explosion
90	Flare	1) Flammable Liquids and Gas (Natural Gas, hydrocarbons liquids)	Jet Fire, Pool Fire, Flash Fires, Explosion
72	Makeup Storage	1) Flammable Gas (Natural Gas) - blanket 2) Flammable Liquid (Methanol Combustible liquids (Amine, TGE, Lube Oil)	Jet Fire, Pool Fire, Flash Fires
BUILDINGS (BI)			
40	Chromatic Raft Building (Lab)	No Information Available - HOLD	
61	ECC Building	No Information Available - HOLD	
63	Emergency Generator Building	1) Combustible Liquid (Lube Oil)	Pool Fire, Spray Fire
CR	Control Room	1) Combustible Solids (paper, electronics, furniture)	
General Area			
PK	Parking Lot	1) ignition Source 2) limited flammable liquid (gasoline inside car tanks)	Pool Fire, Car Fire

Hazardous Area No.	Area Class: Class 1 Div 1	Class 1 Div 2	Class 2 Div 1	Class 2 Div 2	Group	Non-Classified	Ventilation: Natural	Mechanical	Mechanical Supplementary	Overpressure	Underpressure	Occupancy: Permanently Manned	Normally Manned	Normally Unmanned	No Occupancy	Area: Open	Sheltered	Enclosed	Fire Rated Division	Blast Rated Division	Sound: No/Low Noise	Medium Noise	High Noise
Process Areas (PA)																							
31 & 32		✔			D		✔							✔		✔							
41 & 42		✔			D		✔							✔		✔							
63						✔	✔							✔		✔							
21 & 22		✔			D		✔							✔		✔							
26						✔	✔							✔		✔							
91						✔	✔							✔		✔							
14		✔			D		✔							✔		✔							
40		✔			D		✔							✔		✔							
50		✔			D		✔							✔		✔							
80		✔			D		✔							✔		✔							
70		✔			D		✔							✔		✔							
10a		✔			D		✔							✔		✔							
10b	✔	✔			D		✔							✔		✔							
11 & 12		✔			D		✔							✔		✔							
71 & 92	✔	✔			D		✔							✔		✔							
73		✔			D		✔							✔		✔							
90						✔	✔							✔		✔							
72		✔			D		✔							✔		✔							
BUILDINGS (BI)																							
40						✔								✔				✔					
61						✔		✔						✔				✔					
63						✔		✔															
CR						✔		✔				✔						✔			✔		
General Area																							
PK						✔	✔					✔									✔		

Hazardous Area No.	Fire Protection: Deluge/sprinkler (Water)	Foam Suppression System	Gas Suppression System (e.g. CO2, Clean Gas)	Monitor (See note 2)	Hydrant (Water)	Hose Reel (Water)	Portable extinguisher (Powder)	Portable Extinguisher (Foam)	Portable Extinguisher (CO2)	Wheeled Halotron Extinguisher	Wheeled Fire Extinguisher (Powder)	ESD: Manual Emergency Shutdown pushbutton
Process Areas (PA)												
31 & 32							✔			✔		✔
41 & 42												
63										✔		✔
21 & 22												
26												✔
91												
14												
40							✔					
50												
80												
70												
10a												
10b							✔					✔
11 & 12							✔					✔
71 & 92							✔					✔
73							✔					✔
90												
72							✔					
BUILDINGS (BI)												
40												
61										✔		✔
63												
CR							✔			✔		✔
General Area												
PK												

Hazardous Area No.	Detection and Alarms: Pneumatic Heat Detector	Flame Detector (UV / IR)	Rate compensated Heat Detector	Smoke Optical Detector	VESDA Smoke Detection	LOS Flammable Gas Detector	Point Flammable Gas Detector	Flammable Gas Detector (Ducts)	H2S Gas Detector	SO2 Detector	CO2 Detector	Manual Extinguishant Release	Manual Alarm Callpoint	Manual Deluge Release	Extinguishant Status Lamp	Extinguishant pre discharge alarm	Audible alarm (GAPA) Devices (e.g. Horns, Bells)	Visual alarm devices (e.g. beacons)	Local Area Gas Alarm Visual	Supervisory Fire Protection System Alarm	Others: Eye wash station	Emergency showers	First Aid Kit	Wind Sock		
Process Areas (PA)																										
31 & 32								✔													✔			✔		
41 & 42		✔						✔									✔					✔				
63																										
21 & 22								✔																		
26								✔																		
91																										
14																								✔		
40		✔						✔													✔					
50																										
80								✔																✔		
70																										
10a								✔																		
10b																										
11 & 12								✔													✔					
71 & 92								✔																✔		
73								✔																		
90																										
72																					✔			✔		
BUILDINGS (BI)																										
40																										
61																										
63																										
CR				✔																			✔			
General Area																										
PK																										

Figura 2. Ejemplo de diagrama de aras de seguridad (Area Safety Chart en Ingles)

2.3.2 Estimación de Riesgo

El riesgo es la combinación de la severidad de la consecuencia y la probabilidad de ocurrencia de un evento dado. De tal forma que la evaluación de riesgo debe considerar el efecto de la consecuencia del evento y la posibilidad de que el mismo ocurra. El riesgo es una medida que permite la toma de decisiones y priorizar los recursos y controles asegurando una operación segura y controlada.

Es importante entender que diseñar sistemas de protección considerando el peor caso posible, en muchos casos, no es práctico y muy costoso; de forma que la estimación del riesgo permite establecer un balance entre el efecto del evento y la probabilidad de ocurrencia optimizando la relación costo-beneficio.

Por ejemplo, si consideramos un tanque de producto inflamable, el fallo catastrófico del tanque liberando todo su contenido generaría una severidad de consecuencia muy alta, sin embargo, la probabilidad de que esto ocurra es muy baja. Por otro lado, la probabilidad de ruptura de su conexión de alimentación posee una probabilidad mucho más alta, aunque una severidad menor. Considerando el primer evento, digamos que su severidad podría generar 100 fatalidades, pero con una frecuencia relativa de ocurrencia de 1.00 E-08/año, esto generaría un riesgo esperado de 1.00 E-06 fatalidades/año (100 fatalidades x 1E-08/año). El segundo evento podría generar 10 fatalidades, pero a una frecuencia relativa de ocurrencia de 1.00 E-04/año, lo cual genera un riesgo esperado de 1.00 E-03 fatalidades/año (10 fatalidades x 1E-04/año). Se puede observar que el riesgo del segundo evento es mayor por lo que la consideración de dicho evento en los sistemas de protección es necesario. Esta evaluación debe incluir la comparación con un nivel de riesgo tolerable, el cual es valor umbral de riesgo que la empresa u organización está dispuesta a tolerar. De tal forma que la consideración de cada evento dependerá del nivel de tolerancia al riesgo definido; por ejemplo, con un nivel de tolerancia de 1.00 E-04 fatalidades/año el evento catastrófico posee un riesgo bajo y produciría un costo mucho mayor al ser considerado en el diseño, de tal forma que no es típicamente considerado.

El análisis de riesgo en el FHA es utilizado en la selección de incidentes potenciales que deben ser incluidos en el estudio. Aunque la frecuencia es un factor importante durante el análisis de riesgo, desde el punto de vista de protección contra incendios, nos interesa conocer en detalle la severidad de las consecuencias, debido a que, estos sistemas son típicamente mitigativos, utilizados una vez ocurrido el evento. De tal modo que el análisis de riesgo debe incluir un análisis de consecuencias usando herramientas computacionales, de forma que los efectos potenciales de los eventos a considerar sean entendidos y considerados en la respuesta y el diseño de sistemas de protección.

El análisis de riesgo se usa como la base para la selección de los eventos a considerar en el cálculo de consecuencias. La Figura 3 muestra un ejemplo de cálculo de consecuencias mencionado, en dicho ejemplo se muestra los efectos térmicos de un incendio de chorro (jet fire).

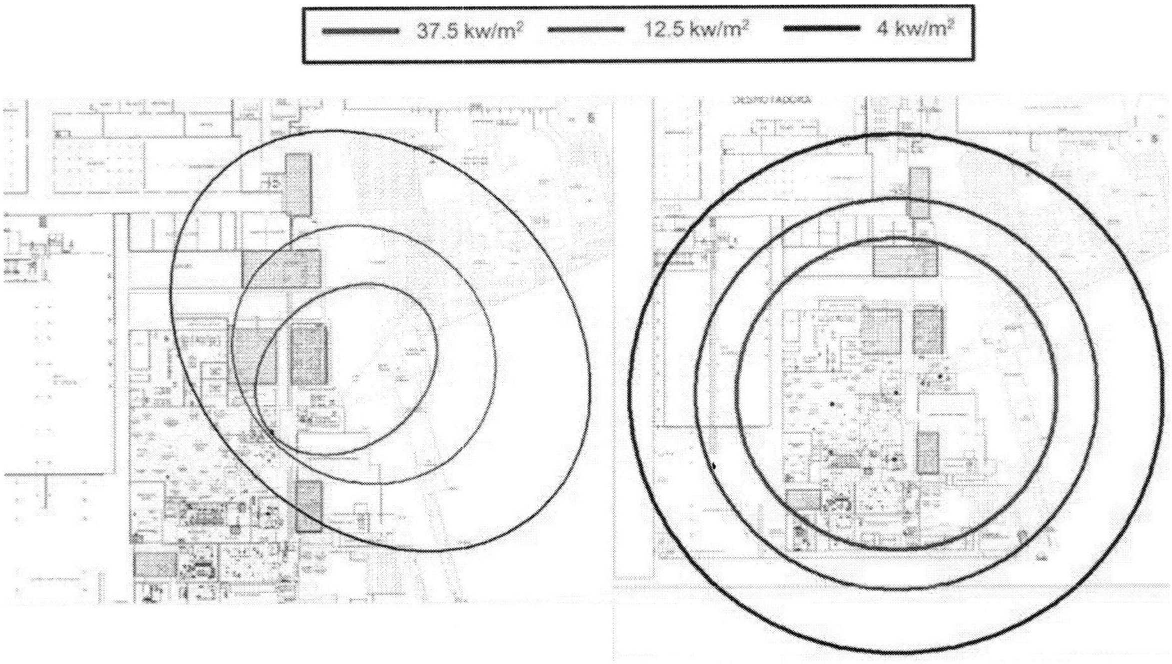

Figura 3. Ejemplo de análisis de consecuencias

El objetivo primordial de la evaluación de riesgo es la definición de los eventos a modelar y que serán considerados en el FHA. El modelamiento de dichos eventos permite entender las consecuencias que pueden generar en caso de que ocurran. Existen tres tipos de evaluación de riesgo que se puede considerar en este caso:

1. Análisis cualitativo.
2. Índice de riesgo.
3. Análisis Semicuantitativo.
4. Análisis cuantitativo.

La metodología de estimación de riesgo usada dependerá del objetivo definido en la primera etapa, por ejemplo, objetivos complejos necesitarían un análisis cuantitativo mientras objetivos menos complejos se podrán cumplir con un método cualitativo utilizando una matriz de riesgo.

2.3.2.1 Análisis de riesgo cualitativo

Este tipo de análisis es basado en el juicio y criterio de los participantes del análisis. El riesgo es subjetivo y es estimado utilizando una matriz de riesgo, la cual es definida antes de realizar el estudio. Muchas organizaciones definen la matriz de riesgo corporativamente para asegurar una continuidad y repetibilidad de los estudios. La Figura 4 muestra un ejemplo de una matriz de riesgo. Sin embargo, las especificaciones de la matriz pueden cambiar dependiendo de la organización.

Este tipo de evaluación de riesgo es genérica y debe ser utilizada como primera instancia durante el taller de identificación de peligros. Desde el punto de vista de protección contra incendio, este tipo de análisis es incompleto ya que, aunque califica la severidad de los eventos potenciales dentro de una categoría, no permite la evaluación de la filosofía de protección ni permite la priorización del tamaño del incidente, p.ej. tamaño de fuga. Sin embargo, el método permite realizar una primera priorización indicando cuales eventos deben incluirse en el FHA y ser evaluados con un método más detallado. Por ejemplo, después de realizar la identificación de peligros y evaluar el riesgo cualitativamente, se puede determinar algunos eventos con riesgo bajo, los cuales no tienen porque ser incluidos en el FHA. Sin embargo, otros eventos como fuga de materiales inflamables y reacciones desbocadas (runaway reactions) requieren un análisis más detallado.

						Probabilidad de Ocurrencia			
						Muy Probable	Probable	Improbable	Remota
		Personas	Daños	Ambiente	Reputación	Ha ocurrido en la instlación	Ha ocurrido en la organización	Ha ocurrido en la industria	No ha ocurrido en la industria
Severidad	Catastrófica	Fatalidad o discapacidad permanente	Daño >$250K	Exceso de cantidad reportable con impacto ambiental mayor	Impacto negativo a la comunidad local con involucramiento de agentes gubernamentales	Alto	Alto	Alto	Mediano
	Seria	Lession seria o enfermidad ocupacional	$100K < Daño < $250K	Exceso de cantidad reportable con impacto ambiental	Violacion encontrada por gobierno local con minimo impacto a la comunidad	Alto	Alto	Mediano	Bajo
	Moderada	Lessión o enfermedad que requiere tratamiento médico	$10K < Daño < $100K	Catidad reportable con impacto ambiental minimo	Notificacion de gobierno local	Alto	Mediano	Bajo	Despreciable
	Leve	Primeros auxilios sin reposo	Daño<$10,000	Fuga por debajo de la cantidad reportable	No impacto	Mediano	Bajo	Despreciable	Despreciable

Figura 4. Ejemplo de matriz de riesgo

2.3.2.2 Índice de riesgo

Los índices de riesgo son técnicas que proporcionan una medida de riesgo relativa la cual es usada para priorización. Desde el punto de vista de procesos, el índice más usado es el Fire and Explosion Index (F&EI) [4] desarrollados por Dow Chemical Company. Estos índices consideran las características de los procesos y los inventarios para determinar el valor del índice y distancia de impacto.

Para las industrias de manufactura que no posean equipos de procesos químicos, existen otros tipos de índices para el riesgo de incendios que pueden utilizarse, por ejemplo, el método de Gretener descrito en el manual de ingeniería se seguridad contra incendios [5].

Estos tipos de análisis nos permiten medir los diferentes riesgos, clasificarlos e identificar cuales eventos deben ser evaluados con el análisis de consecuencias; sin embargo, una desventaja es que no permite la definición del tamaño de emergencia a modelar ya que en algunos casos la consideración del peor caso sería impráctico y costoso.

2.3.2.3 Análisis de riesgo semicuantitativo

El análisis de riesgo semicuantitativo considera la frecuencia del escenario de forma cuantitativa, pero define la severidad de forma cualitativa asignándola en una categoría predefinida.

Para utilizar este método en toma de decisiones, cada categoría de severidad debe poseer un criterio de tolerabilidad. Típicamente este tipo de análisis compara la estimación de la frecuencia del evento a considerar con el criterio de tolerabilidad. Este valor puede ser considerado para definir si alguna medida adicional de control de riesgo debe considerarse. Desde el punto de vista de protección contra incendio, la estimación de frecuencia solo considera la frecuencia de ocurrencia del evento sin considerar la implementación de sistemas de protección contra incendios, luego se debe definir si los sistemas de protección contra incendios a implementar efectivamente reducen el riesgo a niveles por debajo del criterio de tolerabilidad.

La estimación de frecuencia puede incluir diversas metodologías, p.ej.:

- Análisis de árbol de fallos (FTA por sus siglas en ingles).
- Análisis de árbol de eventos (ETA por sus siglas en ingles).
- Análisis de cadenas de Markov.
- Análisis Monte Carlo.
- Análisis probabilísticos de estadísticas de accidentes.

Desde el punto de vista del FHA, solo se considerarán en el taller los eventos que excedan el criterio tolerabilidad por evento. La Figura 5 muestra un ejemplo de frecuencia de eventos.

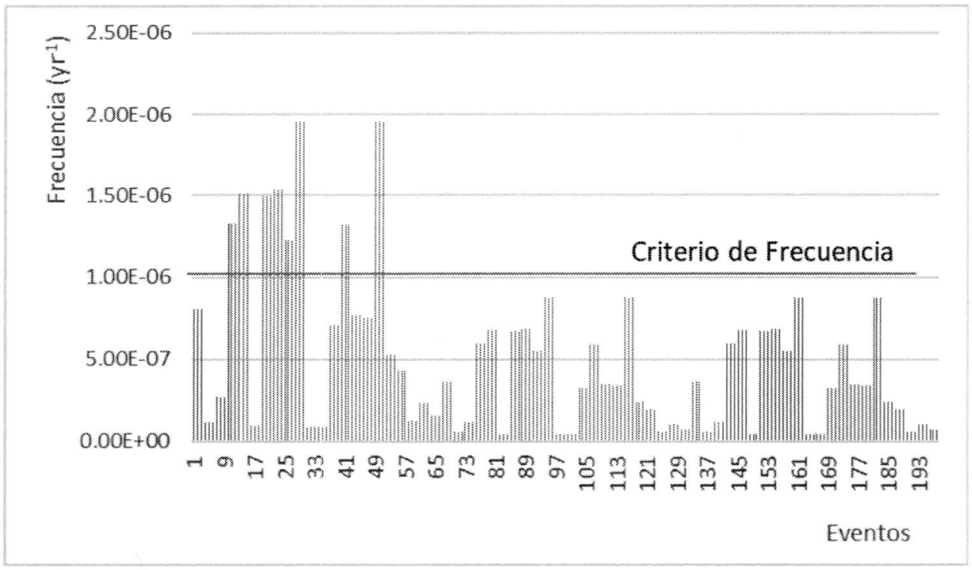

Figura 5. Ejemplo de criterio de frecuencia

2.3.2.4 Análisis de riesgo cuantitativo

El análisis de riesgo cuantitativo considera tanto la severidad como la frecuencia de forma cuantitativa. Existe diversas medidas de riesgo cuantitativas, las cuales son usadas para diversos casos, desde el punto de vista de la protección contra incendios se debe considerar la contribución de cada evento a la frecuencia total y considerar un criterio predefinido para incluir el evento dentro del estudio. El análisis de riesgo cuantitativo típicamente utiliza la frecuencia de fatalidad como medida de riesgo y tomas de decisiones, observe que en este caso particular de definición de medidas de protección contra incendio se utiliza la frecuencia de ocurrencia del evento sin considerar la probabilidad de fatalidad, esto se debe a que la protección contra incendio también busca minimizar la perdida monetaria e impacto a la continuidad operativa.

El criterio de frecuencia de excedencia es usado en este caso. La frecuencia de excedencia es la frecuencia cumulativa de exceder un cierto valor físico. Se plantea usar un criterio de frecuencia de excedencia basada en el área total afectada. Este criterio se basa en que, a mayor área de afectación, la probabilidad de impacto al personal aumenta (lesiones o fatalidades), así como el impacto financiero y área necesaria a proteger. De igual modo es más fácil evaluar la capacidad de los sistemas a instalar considerando el área afectada. Este criterio es más complejo de aplicar ya que requiere la estimación del área de afectación y la evaluación de la curva de excedencia de frecuencia de excedencia-área. La Figura 6 muestra un ejemplo de esta curva con el criterio de excedencia.

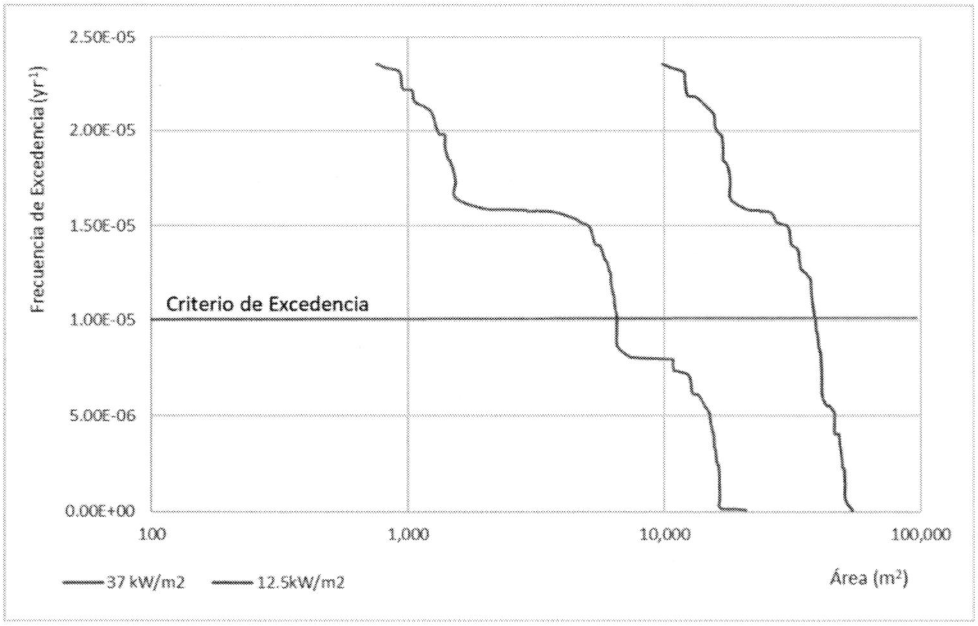

Figura 6. Ejemplo de curva de excedencia

La metodología general de análisis de riesgo cuantitativo incluye los siguientes pasos:

1. Definir condiciones de los procesos.
2. Agrupar secciones con condiciones similares y fuentes de fuga.
3. Cálculo de tasa de fuga.
4. Estimación de probabilidad de ignición.
5. Cálculo de consecuencias de los eventos y estimación de área de afectación usando los contornos de efectos (p.ej., radiación térmica).
6. Conteo de partes clasificados por fuentes de fuga.
7. Análisis de frecuencia clasificado por fuente de fuga.
8. Estimación de frecuencia de incendios, explosiones y nube toxica.
9. Estimación de frecuencia cumulativa de excedencia de áreas clasificados por grupo de fugas y tipo de incidente.
10. Estimación de riesgo de cada evento.
11. Identificación de eventos que cumpla con el criterio de tolerabilidad.
12. Seleccionar eventos representativos por grupo de fuga a ser considerados como base de la respuesta de emergencia interna.

A pesar de que este método es más analítico, tiene la desventaja que debe realizarse el cálculo de consecuencia de todos los escenarios previstos para su aplicación. Sin embargo, esto no indica que todos serán analizados en el taller, el análisis indicará cuales serán analizados en el taller FHA.

2.3.3 Análisis de Consecuencias

Durante esta etapa se toma los peligros y eventos descritos en el diagrama de áreas de incendios, descrito en la sección anterior (2.3.1.1), y se realiza una modelación o simulación con el fin de determinar los efectos producidos por dicho incendio. Sin embargo, debido a que el diagrama de áreas de incendios puede incluir una cantidad significante de eventos, debe realizarse una revisión de estos para determinar cuáles eventos son más significativos. De la misma forma, debe definirse los parámetros importantes a considerarse en el modelo. Al realizar esta revisión, los escenarios de incendios a modelar serán definidos.

Vale la pena indicar la diferencia entre evento y escenario; los eventos, son definidos de forma genérica indicando la fuente de combustible y lugar de ocurrencia, p.ej., incendio generado por fuga de propano en la columna absorción T-001, ubicada en la planta fraccionadora PF-002. Los escenarios son incendios específicos asociados a un evento; los mismos son definidos en detalle proporcionando información de las variables importantes. Estos parámetros incluyen:

- Evento asociado.
- Tamaño de ruptura.
- Tipo de incendio (charco, chorro, destello).
- Dirección de fuga.
- Clima.

  - Dirección de viento.
  - Velocidad de viento.
  - Estabilidad atmosférica.
  - Temperatura atmosférica.
  - Humedad relativa.
- Información asociada a la operación y/o respuesta:
  - Turno de trabajo (noche, día, almuerzo, etc.).
  - Activación de sistemas de aislamiento o control de fugas, p.ej.:
    - Antes o después de activación de aislamiento.
    - Falla o activación de válvulas de aislamiento de emergencia.
    - Falla o activación de sistema de despresurización.
  - Operación realizada. p.ej., en operaciones tipo batch.

Es importante destacar que cada evento de incendio puede tener un sinfín de escenarios asociados, por tal motivo que los escenarios a modelar deben ser escogidos adecuadamente de forma que estos representen el peor caso. Para esto se definen los llamados escenarios de diseño o escenarios representativos, los cuales son escenarios que pueden ser controlados de forma practica con sistemas de protección contra incendios. De forma que estos escenarios no representan el peor caso posible, pero tampoco el caso mas leve.

Para eventos de perdida de contención, el cálculo de consecuencias requiere la definición del tamaño de la fuga, lamentablemente no existe un consenso en la industria respecto al tamaño a usar [8], la guía de análisis de consecuencias de CCPS [8] sugiere un enfoque para la selección del tamaño de fuga, pero el mismo es específico para análisis de riesgo y no define el tamaño a considerar para diseño de sistemas de protección. En tal sentido, la mejor opción analítica sería realizar un análisis semicuantitativo o cuantitativo de riesgo para escoger los escenarios. En caso de realizar el análisis cuantitativo, se utilizarían diversos tamaños, pero para un análisis semicuantitativo se escogería la fuga que exceda el criterio de tolerabilidad. Sin embargo, algunas organizaciones han definido los siguientes criterios:

Si Diámetro de conexión < 51 mm (2 in) se modela ruptura total.

Si Diámetro de conexión > 51 mm (2 in) se modela una fuga de 51 mm.

Estos valores han sido considerados ya que generalmente en estudios cuantitativos de riesgo, la fuga de 50 mm ha producido un valor de excedencia de al menos 90%. Esto significa que la frecuencia acumulada de todos los incendios producidos por fugas menores de 50 mm representa el 90% de todas las fugas. Por lo tanto, a pesar de que fugas mayores a 50 mm producirían un incendio mucho más severo y devastador, poseen una probabilidad de ocurrencia mucho más baja y el diseño de un sistema de protección para estos escenarios serían muy costos y poco prácticos. Esta observación es genérica, y cada caso en particular podría ser diferente. Otro factor que considerar es que fugas mayores tendrán una duración de fuga menor ya que poseen un flujo de fuga más alto, a diferencia de las fugas menores que poseen un flujo de fuga menor produciendo una duración de fuga mucho mayor.

Para eventos que no requieran una pérdida de contención, como reacciones desbocadas, la base del modelamiento es evaluada en base a las condiciones normales de operación y las fallas potenciales definidas en análisis de peligros de los procesos como HAZOP. Por ejemplo, una reacción exotérmica desbocada de un reactor puede ocurrir debido a la falla del sistema de enfriamiento, el modelamiento de la consecuencia se realizaría considerando las cantidades de reactivos normales de operación para estimar el incremento de presión y temperatura y posible explosión interna del reactor o generando de material inflamable; también se debería evaluar la apertura de válvulas de alivio si las mismas liberan material al ambiente. La inclusión de estos eventos en análisis semicuantitativos o cuantitativos puede realizarse al estimar la frecuencia de dichos eventos utilizando análisis de árboles de falla (FTA).

Existen otros factores a considerar durante la definición de los escenarios de diseño, estos normalmente consideran los parámetros que generen el peor caso posible. De forma que se pude establecer el escenario de diseño como el peor caso posible resultante de la fuga de diseño ya anteriormente definida.

Una vez se establezca la lista escenarios de diseño, se debe realizar los cálculos de consecuencias estimando los siguientes parámetros:

- Dimensiones de llama.
- Contornos de flujo de calor por radiación significativos. típicamente se usan los siguientes:
  - 37.5 kW/m^2 – flujo de calor mínimo para producir fallo del acero. Producción de escalación y BLEVE al exponer contenedores con material inflamable por 10 minutos.
  - 12.5 kW/m^2 – flujo mínimo de ignición de madera y material a base de celulosa. Produce quemaduras de 1ro y 2do grado en menos de 10 segundos sin protección alguna, se asume como el flujo máximo de calor soportado por un respondedor utilizando el equipo de protección personal de combate de incendio (traje de bombero) por un tiempo limitado.
  - 6.3 kW/m^2 – flujo de calor máximo soportado por personas durante un minuto de exposición utilizando ropa adecuada. Limita acciones de emergencia con una duración de un minuto. Personal de planta no podría acceder a áreas expuestas a un flujo mayor.

- Diagramas de los dimensiones y contornos a escala contrapuestos a planos de la instalación.

Los resultados obtenidos por el calculo de consecuencia para cada escenario de diseño representa la base del estudio y será revisada por un equipo multidisciplinario en el taller FHA.

2.3.3.1 Características del modelo de consecuencias

Debido a que el cálculo de consecuencia representa la base del estudio, es de vital importancia que estos cálculos sean adecuados. Por lo tanto, que la escogencia del modelo posee la capacidad de afectar los resultados del estudio, de forma que es importante que se escoja el modelo mas adecuado.

Los modelos de consecuencia de incendios deben poseer dos cualidades importantes:

1. Sean matemáticamente coherentes.
2. Representen con un grado aceptable de incertidumbre la realidad del fenómeno de incendio.

De tal forma que el modelo a utilizar debe ser propiamente Validado y Verificado. Este requerimiento es típicamente realizado por los desarrolladores de los softwares y modelo; sin embargo, el usuario debe realizar la debida diligencia de asegurarse que el modelo utilizado haya pasado exitosamente por un proceso de verificación y validación. En estados unidos la Sociedad Americana de Pruebas y Materiales (ASTM por sus siglas en inglés) describe los requerimientos de estos procesos en su es estándar ASTM E1355 Guía para la evaluación de la capacidad prescriptiva de los modelos determinísticos de incendios. Este estándar define dichos procesos de la siguiente manera:

- Verificación: Proceso para determinar que la implementación del método de cálculo representa precisamente la descripción conceptual del desarrollador sobre el método de cálculo y su solución.
- Validación: El proceso de determinar el grado el cual el método de cálculo es una representación precisa del mundo real desde la perspectiva del uso destinado del método de cálculo.

En palabras más sencillas, la verificación busca asegurar que las ecuaciones matemáticas sean implementadas adecuadamente, y la validación busca asegurar que el modelo haga una predicción adecuada con una incertidumbre apropiada para el uso que se le va a dar. La validación normalmente se realiza comparando la predicción del modelo con resultados experimentales, mientras la verificación compara parámetros de diversos casos para entender la repetibilidad y consistencia de los resultados.

Desde el punto de vista del usuario del modelo, las siguientes actividades podrán ser realizadas para asegurar que el modelo usado es adecuado [7]:

- Solicitar y leer informe de validación y verificación a desarrolladores.

- Correr casos simples donde el resultado sea fácilmente predecible y compararlo con los resultados del modelo. Se busca casos que sean fácil de calcular manualmente y comparar resultados. Variar diversos parámetros para verificar consistencia de los resultados.

- Correr análisis de sensibilidad de un caso en particular. Verificar si el resultado obtenido corresponde a lo esperado; p.ej., incrementar tamaño de fuga aumenta flujo de escape, dimensiones de llama, contornos de radiación, y disminución de duración de fuga.

- Correr análisis de caso particular. Verificar si el resultado tiene sentido físico; p.ej., tiene sentido el tamaño de llama de este caso, tiene sentido la velocidad de descarga, etc.

- Validación indirecta. Este tipo de validación es cuando se compara los resultados del modelo evaluado con los resultados de otro modelo que haya sido aceptado previamente para un uso similar; p.ej., correr el mismo caso y comparar el modelo evaluado con resultados obtenidos por un modelo de fluido dinámica computacional (CFD por sus siglas en ingles). Hay que entender, que habrá una diferencia entre modelos, lo importante es entender si la diferencia puede impactar el resultado significativamente.

Una vez se hayan realizados las debidas diligencias asegurando que los modelos a utilizar hayan sido debidamente verificados y validados para el uso deseado, se pueden aplicar para predecir los efectos de calor y dimensiones de llama de los escenarios de diseño definidos.

2.4 Evaluación de desempeño

Antes de empezar el taller FHA, se debe evaluar el desempeño de los sistemas de protección que se encuentren instalados o diseñados. Esta evaluación ayudara a analizar su efectividad durante el taller, de igual forma permite recomendar mejoras que aumenten la efectividad de los mismos.

Desde un punto de vista práctico, esta evaluación consiste en la revisión del diseño del sistema, comparándola con las mejores practicas en la industria; la cual puede incluir calculo hidráulicos de sistemas a base de agua, capacidad de agente extinguidor, flujos de descarga, capacidad de diques de contención secundaria, etc.

2.5 Taller FHA

El taller FHA es una reunión multidisciplinaria donde se evalúan los diversos escenarios de incendios identificados definiendo la manera óptima y práctica de prevenir, controlar o mitigar el evento. Para considerar todos los variables presentes en cada evento, el taller debe incluir personal de operación, seguridad, respuesta de emergencias y disciplinas importantes de ingeniería (por ejemplo, procesos y automatización).

La dinámica del taller incluye una revisión detallada de cada escenario desde el momento de su ocurrencia hasta su respuesta. La figura 7 muestra las diferentes etapas de discusión realizada por cada escenario de incendio.

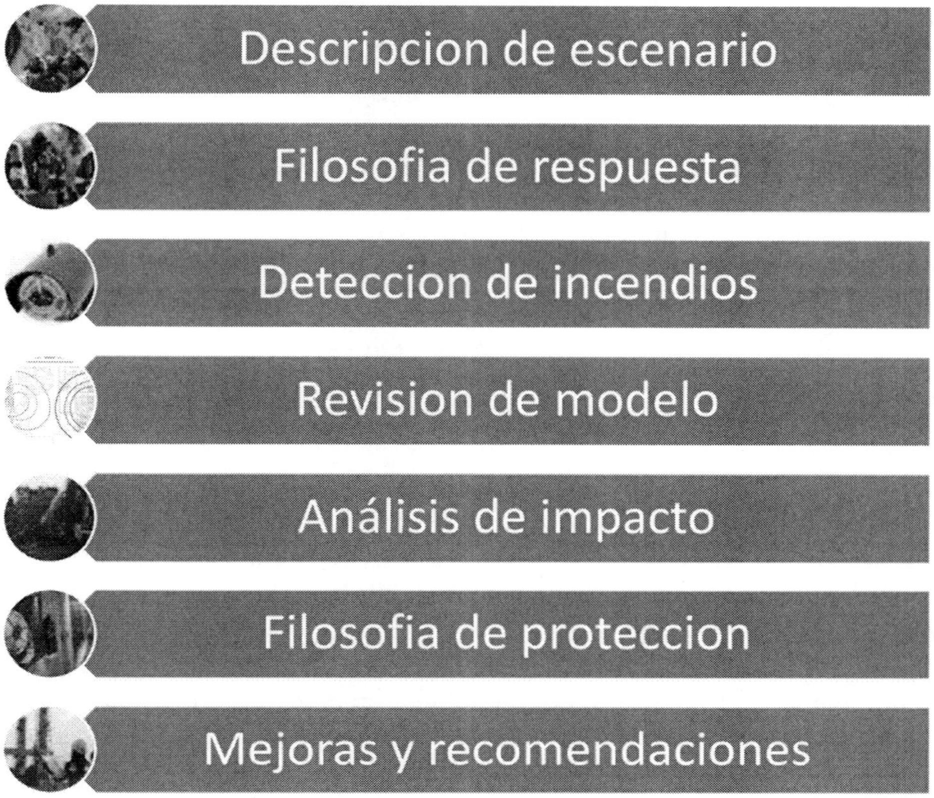

Figura 7. Metodología general de evaluación de escenario en taller FHA

2.5.1 Descripción de escenario

Lo primero que debe hacerse es la descripción del escenario a discutir, la idea es que las personas involucradas en el taller entiendan el escenario y posean la información mínima necesaria para poder analizarlo y proveer información útil. La descripción debe contener al menos la siguiente información:

- Número de identificación del escenario.
- Lugar de ocurrencia.
- Material involucrado.
- Tipo de incendio, p.ej., charco, chorro.
- Clima, p.ej., velocidad de viento, dirección del viento, humedad relativa, etc.
- Momento de ocurrencia, p.ej., turno de trabajo, operación especifica.
- Características o condiciones durante el evento.

2.5.2 Filosofía de Respuesta

En esta etapa se define la filosofía definida para responder y controlar el incendio de diseño estudiado. Durante esta etapa no se realiza ningún juicio sobre la efectividad de la filosofía. La idea es plantear el escenario y la respuesta como base de la discusión.

La descripción de la filosofía de respuesta debe incluir la siguiente información:

- Sistemas implementados para el control.
    - Modo de activación, p.ej., manual, automático.
    - Lugar de activación, p.ej., cuarto de control, local, etc.
    - Agente extinguidor en casos de sistemas activos de supresión, p.ej., agua, espuma, etc.
    - Personal que debe operar el sistema, p.ej., bomberos de la zona, personal de planta, etc.
- Acciones realizadas por personal de planta.
- Acciones esperadas para personal no afectada, p.ej., evacuación, etc.
- Movimiento de equipos y recursos necesarios y tiempo estimado para dicho movimiento, p.ej., movimiento de vehículos de combate de incendio.
- Autoridad durante la emergencia.

De la misma manera, esta descripción también incluye la información obtenida durante la evaluación de desempeño de los sistemas de protección. Cuando esta evaluación es base a un modelo, p.ej., calculo hidráulico, el mismo será revisado mostrando los datos de entrada y los resultados obtenidos.

2.5.3 Detección de incendios

La detección de incendio es el mecanismo por la cual se identifica la presencia de un incendio y se avisa al personal de planta y respuesta su ocurrencia. El método utilizado para la detección y alarma es fundamental en el control del incendio, ya que permite la activación temprana de los medios de respuestas disminuyendo su potencial impacto.

La detección de incendios puede realizarse por medio de elementos tecnológicos o humanos. En el medio tecnológico elementos sensores responden a los efectos producidos por el incendio (p.ej., calor, luz infra roja, etc.) proporcionando una señal indicando su presencia. Por lo contrario, una detección humana se basa en la identificación del evento por personal de planta durante sus rondas de vigilancia y revisión.

La primera discusión realizada en el taller es sobre el medio implementado para la detección del evento y la definición del posible tiempo transcurrido entre la ocurrencia del evento y el momento en que el operador toma una acción. Este tiempo incluye diversos tiempos mostrados en la figura 8.

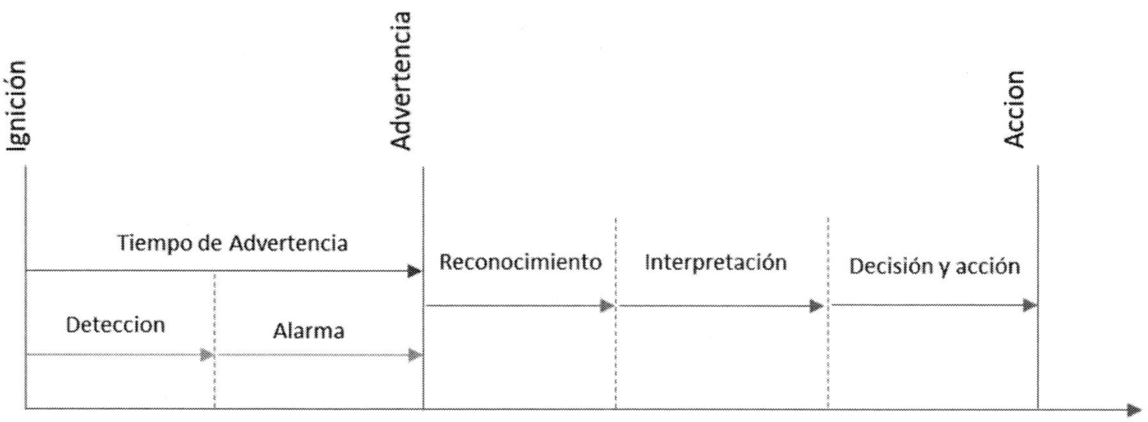

Figura 8. Tiempo para acción

Al entender estos tiempos, se puede entender que el tiempo el cual el incendio estará ardiendo sin control y el momento que se empieza a aplicar alguna medida de control. Es importante que estos tiempos se definan numéricamente, ya que serán necesarios en el análisis de impacto.

2.5.4 Revisión del modelo

Una vez se haya descrito las bases del escenario a discutir, se procede a mostrar el cálculo de consecuencia realizado de dicho escenario. Sin indagar en mucho detalle técnico, se muestran los parámetros de entrada, y los resultados obtenidos.

En casos de pérdidas de contención se empieza mostrando el modelo de descarga transitorio mostrando el flujo de descarga versus el tiempo. Se observa el tiempo total de descarga y se analiza como dicha curva interactúa con los tiempos de respuestas y de detección descritos anteriormente para definir un tiempo de duración del escenario de incendio. P.ej., el tiempo de duración de un incendio tipo chorro incluirá el tiempo de descarga una vez aislado el sistema más el tiempo transcurrido desde el momento de la ignición hasta el tiempo que las válvulas de aislamientos se encuentran cerradas.

Una vez conocido la duración del evento, se procede a mostrar los resultados de la dimensión de llama y flujos de radiación. Los diagramas deberían sobreponerse a plano de la instalación para poder entender el impacto a diversos tiempos. Esta revisión será la base del análisis de impacto.

2.5.5 Análisis de impacto

El análisis de impacto comprende la revisión de los resultados del cálculo de consecuencia conjuntamente con los tiempos definidos para analizar el posible impacto que podría generar el escenario discutido. Durante esta discusión se evalúa la efectividad que podría tener los sistemas de control de incendios definidos para el evento y si alguna mejora se podría realizarse. Durante esta discusión, podrían realizarse modificaciones del cálculo de consecuencia para entender el impacto que podría tener alguna mejora o contexto no incluido en el modelo.

El impacto no solo debe incluir el potencial fallo de los materiales de construcción y estructuras, sino cual efecto podría resultar si estos fallan. P.ej., si se define que una columna falla térmicamente por efectos del incendio, podría colapsar el edificio y afectar negativamente a los ocupantes, o que efecto tendría en la continuidad del negocio que colapse dicho edificio?

2.5.6 Filosofía de protección

Una vez analizado el impacto, se juzga la filosofía de protección implementada indicando si es adecuada. De la misma forma, se realizan modificaciones para asegurar el control y protección.

2.5.7 Mejoras y recomendaciones

El fin del taller es evaluar los sistemas de control y protección de incendio, por lo tanto, los participantes del taller deben realizar recomendación de mejoras a la filosofía o diseño del sistema. Todas estas recomendaciones deben ser propiamente documentadas como resultado del estudio FHA.

3 Caso de estudio

Como caso de estudio vamos a considerar una instalación de almacenamiento y distribución de hidrocarburos. La instalación tiene la capacidad de cargar tanques atmosféricos a través de tuberías y de procesos de descarga de buques. Por otro lado, la instalación tiene la capacidad de enviar material desde los tanques atmosféricos a la estación de carga de camiones cisterna, donde llenan los mismos y los despachan para distribuir el hidrocarburo. La figura 9 muestra la distribución general de la instalación.

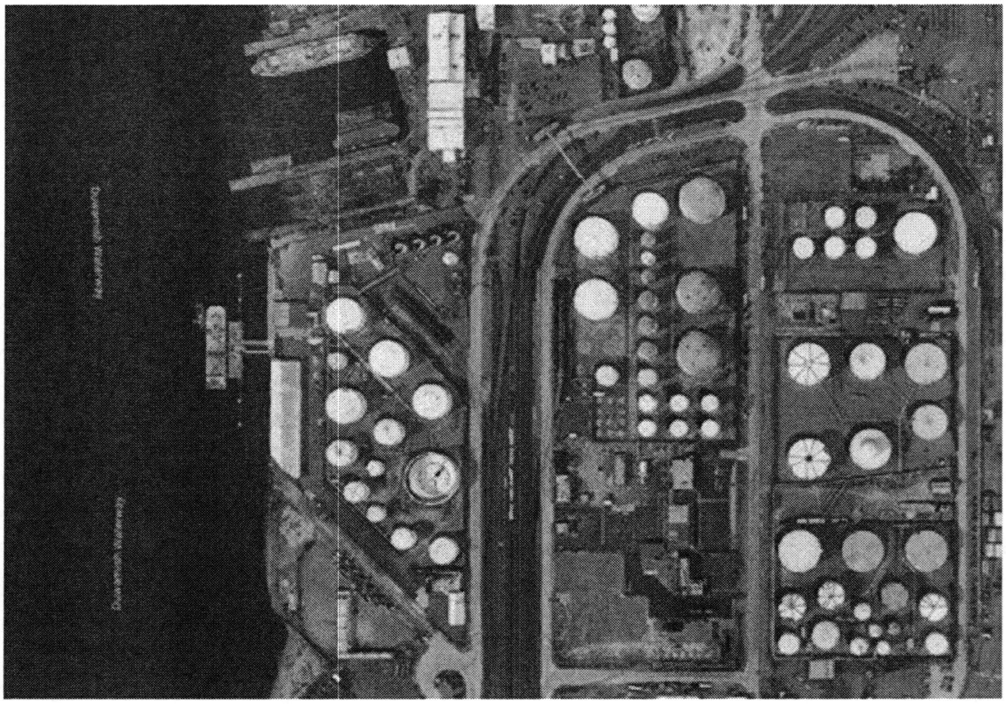

Figura 9. Terminal de carga y descarga

El terminal ha presentado problemas de desempeño del sistema de protección contra incendio que protege el área de carga de camiones cisterna. Por tal motivo, que la organización desea entender las bases de protección buscando una forma óptima de proteger dicha área. Por tal motivo se decidió realizar un FHA para analizar la situación.

3.1 Objetivos y alcance

Para este caso especifico el objetivo del FHA es la identificación de mejoras del sistema de protección contra incendio en el proceso de carga de combustible. El alcance incluye los incendios ocurridos durante las operaciones de carga de combustible en los camiones cisterna. El alcance incluye solo los incendios ocurridos en el área de carga, no incluye los tanques de almacenamiento, ni el proceso de carga de los tanques. La figura 10 muestra el área de carga y el cuarto de control.

Figura 10. Terminal de carga y descarga

3.2 Contexto

Esta instalación fue construida a finales de los años 70, al momento del estudio tenía aproximadamente 40 años. La instalación ha pasado por diversos dueños durante este tiempo, los cuales realizaron diferentes modificaciones para ajustar el terminal a sus necesidades de negocios.

El terminal original contaba con tres carriles de carga, y a lo largo de los años se han expandido a siete carriles. Sin embargo, al momento del estudio, no todos los carriles eran utilizados constantemente; en promedio existían 3 carriles en uso constante.

Debido al cambio de dueños, no existía documentación de diseño o de mantenimiento del sistema de protección contra incendio; el único historial conocido era las modificaciones y mantenimiento realizado por el dueño actual.

Adicionalmente, el sistema experimentó una activación accidental descargando el contenido completo del tanque de espuma produciendo un impacto ambiental. Razón por la cual se comisionó el mencionado estudio.

3.3 Análisis de riesgo

3.3.1 Identificación de peligros

Se realizó una identificación de peligro revisando el proceso de carga de combustible. Se identificaron diversos escenarios capaces de producir perdidas humanas y materiales.

- Incendios tipo destello. Diversos escenarios resultan en incendio tipo destello (flash fire), los cuales pueden impactar a la seguridad del operador. Sin embargo, la duración de este tipo de eventos es tan corta que el sistema de protección contra incendio no responde a tiempo y el evento no puede ser controlado. Sin embargo, estos tipos de eventos pueden generar incendios tipo piscina posteriores lo cuales pueden ser controlados por el sistema.
- Explosiones. Escenarios con mayor volumen de fuga pueden generar vapores acumulándolos en el área de carga, que al ser congestionada podría generar sobrepresiones cuando los gases encuentren una fuente de ignición. El sistema de protección contra incendio no responde rápidamente y fallaría en controlar estos tipos de eventos.
- Incendio tipo piscina/charco. Estos tipos de incendios son producidos cuando ocurre alguna fuga durante la operación de carga. El flujo liquido se acumula en el piso, y al ser encendidos produce el evento. Este tipo de evento puede ser controlado efectivamente por el sistema de protección contra incendios, sin embargo, el mismo debe estar diseñado para el mismo.
- Incendio tipo chorro. Debido a las presiones manejadas durante la operación de carga, no se identificaron escenarios con estos tipos de incendios.

El evento principal identificado fue un incendio tipo piscina/charco. Aunque diversos escenarios fueron identificados, el escenario de diseño definido fue la fuga debido a la ruptura total de la manguera de carga. Esta manguera de carga tiene una dimensión de 50 mm de diámetro; por lo tanto, cumple con la definición heurística definida en la sección 2.3.3.

Para este caso, como solo se definió un área y un escenario de diseño, no se produjo el documento base de diagrama de áreas de incendios.

3.3.2 Estimación de riesgo

Inicialmente una estimación de riesgo cualitativa basada en la matriz de riesgo corporativa fue utilizada durante la identificación de peligro. Para el FHA se decidió realizar un estimado semicuantitativo, donde la frecuencia del evento es estimada por un árbol de falla y la severidad es definida tal como fue considerada en la etapa de identificación de peligros. Adicionalmente, el dueño de la instalación ha adoptado un criterio de tolerabilidad semicuantitativo (por evento, no cumulativo) el cual fue utilizado para la comparación de riesgo. La Tabla 2 muestra el criterio de tolerabilidad implementado por la organización.

Tabla 2. Ejemplo de tabla de identificación de peligros

Severidad		Criterio Tolerabilidad (yr-1)
0	No lesión o efectos a la salud	0.00E0
1	Lesión o efecto a la salud ligera	1.00 E-01
2	Lesión o efecto a la salud menor	1.00 E-02
3	Lesión o efecto a la salud mayor	1.00 E-03
4	Discapacidad (parcial o total) o hasta 3 fatalidades	1.00 E-04
5	Mas de tres fatalidades	1.00 E-04

Para el escenario de diseño, se consideró una frecuencia de 4 fallas por E06 horas de operación por manguera [11], una probabilidad de ignición (P_{ig}) de 0.03 y una probabilidad de falla a demanda de medidas de ignición de 0.1 (PFD_{ig}).Estos últimos valores fueron definidos por el operador del terminal en sus documentos corporativos de análisis de riesgo, sin embargo la guía del CCPS para determinar la probabilidad de ignición [12] puede ser usada. El valor de la frecuencia de incendio es estimado como la multiplicación de la frecuencia de fuga por las probabilidades asociadas. Considerando el contexto de operación, se asumió una distribución uniforme de uso considerando 3 carriles en uso por hora, resultando en un tiempo de operación de 3.754,3 hr por año por manguera. Al tomar en consideración las 7 mangueras se tiene un uso de 26.280 horas de operación al año. Al considerar todos estos factores se obtiene un valor de frecuencia de incendio de 3,15 E-04. Este valor es mayor al criterio de tolerabilidad de severidad 4 considerado en este caso. Por lo tanto, se necesita el sistema de protección contra incendio para minimizar el riesgo a un valor aceptable.

$$F = IEF \cdot P_{ig} \cdot PFD_{ig}$$

$$\rightarrow F = 26.280\, hr/año \cdot 4.00 \cdot 10^{-6} hr^{-1} \cdot 0.03 \cdot 0.1 = 3.15 \cdot 10^{-4}\, año^{-1}$$

El siguiente paso seria estimar el valor de frecuencia considerando el sistema de protección contra incendio, de forma que se evalúe si se requieren medidas adicionales. Esta evaluación se realizó usando un árbol de fallo del sistema instalado para considerar la probabilidad de falla a demanda (PFD) del sistema. Este árbol produjo un valor de 0,054, generando un valor total de frecuencia de 1,7 E-05. Este valor es menor que el criterio de tolerabilidad demostrando que el sistema de protección seria suficiente para disminuir el riesgo a un valore aceptable de riesgo. La figura 11 muestra el árbol de fallo del sistema de protección.

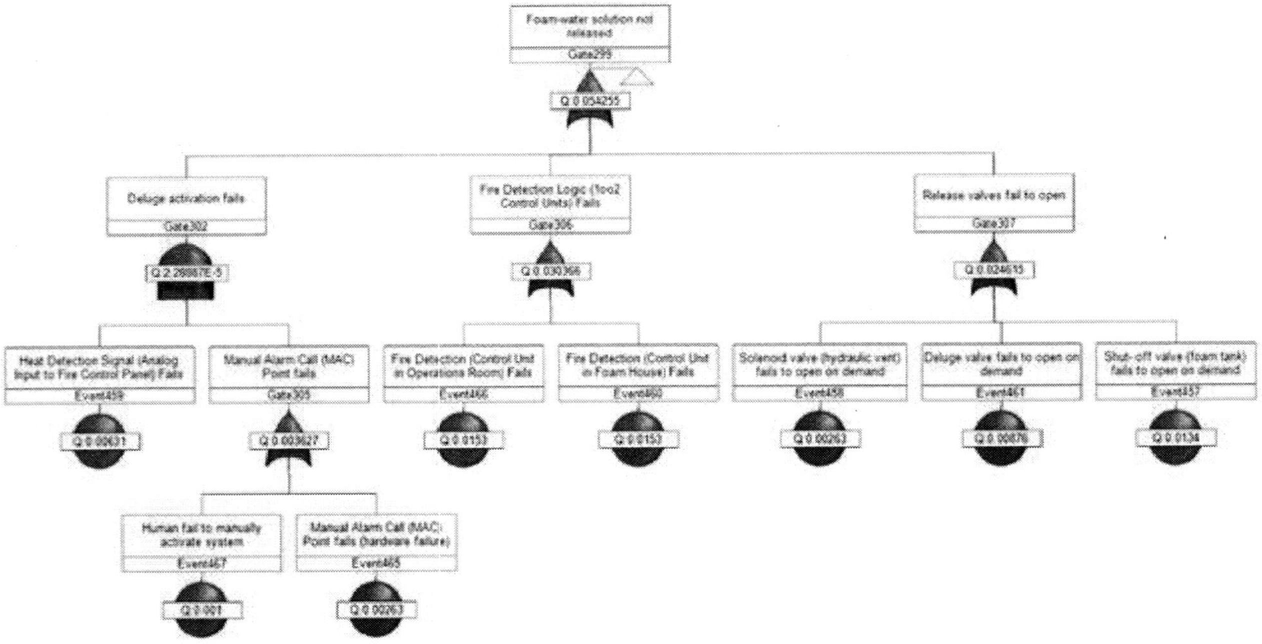

Figura 11. Terminal de carga y descarga

3.3.3 Análisis de consecuencias

Se realizó un cálculo de consecuencia del escenario de diseño para entender el efecto potencial. Se realizaron dos casos, el primero considerando una ignición temprana, la cual se basa en un balance entre la tasa de fuga y la tasa de quemado, generando un tamaño de piscina de 5 mts. El segundo caso, considera una ignición tardía, el cual considera que el material no se encuentra fugando cuando la ignición ocurre, por lo tanto, la piscina crece en tamaño antes de la ignición. Sin embargo, esta última tiene una duración menor. El tamaño de este ultimo caso fue estimado en 14 mts de diámetro.

La figura 12 y 13 muestra el resultado del cálculo de consecuencia de estos casos de diseño.

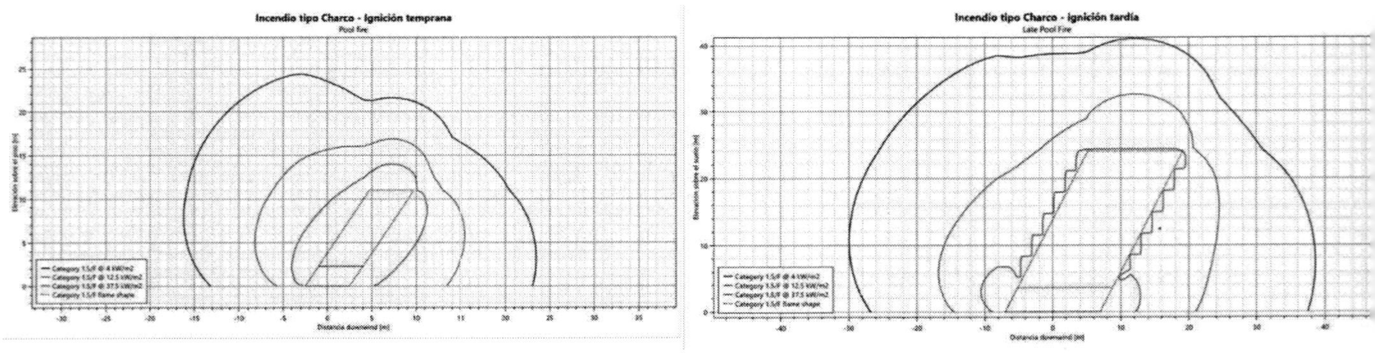

Figura 12. Modelado de escenario - elevación

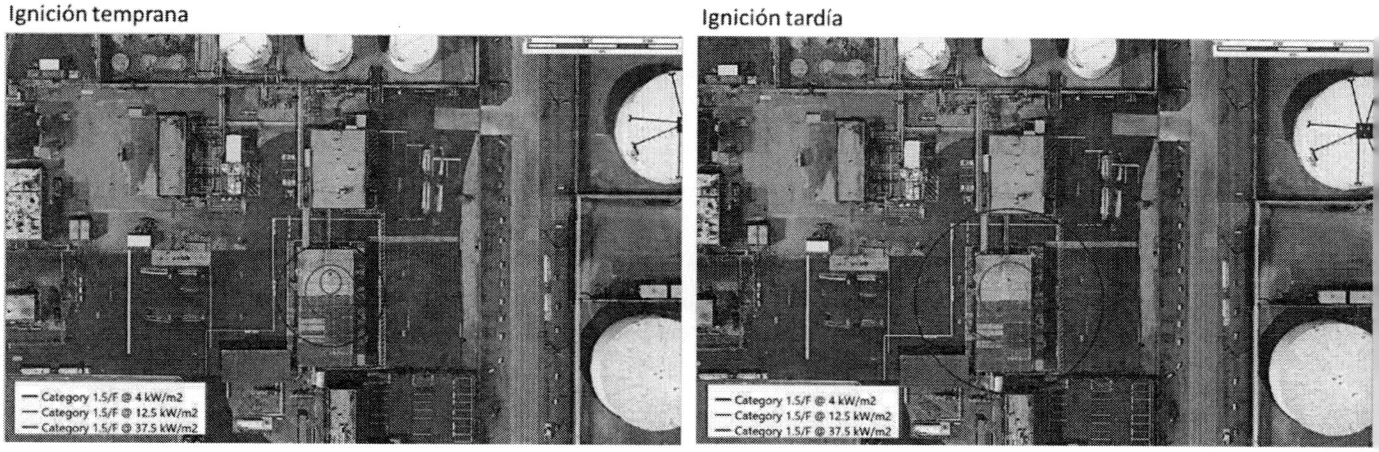

Figura 13. Modelado de escenario – vista aérea

Vale mostrar que la altura de la estructura de donde se realiza las operaciones de carga es de aproximadamente 5 mts, de tal forma que cualquiera de los dos casos produce llamas superiores impactando el techo y esparciéndose lateralmente, de forma que el calor producido divergirá de los contornos producidos por el programa computacional de consecuencia. En realidad, dichos contornos pueden ser mayores, abarcando una mayor área de la sección estudiada.

De tal forma que se realizó un cálculo utilizando el método de fluido dinámica computacional (CFD por sus siglas en ingles) para evaluar las temperaturas, comportamiento de llama y flujo de calor generados por el caso de ignición temprana. La figura 14 muestra la visualización del modelo realizado.

Figura 14. Modelado CFD para caso de ignición temprana

3.4 *Evaluación de desempeño*

La evaluación de desempeño considera el sistema de protección de incendio instalado. Este sistema tomaba agua de la conexión municipal, pasaba por un cuarto de espuma donde el sistema de inducción alimenta el agua con el concentrado para formar una solución de espuma AFFF de 3%. Luego esta línea alimenta a un sistema de rociadores de diluvio que se distribuye en toda el área protegida. El sistema estaba diseñado originalmente para ser activado automáticamente cuando el sistema de alarma detecte un incendio, sin embargo, debido al incidente ocurrido, esta función fue desactivada. Así que el sistema debe ser activado manualmente por un operador que debe abrir la válvula de en el cuarto de espuma. La idea del operador del terminal es que esto puede ser realizado por el personal en el área de carga.

La evaluación de desempeño incluyó una revisión del diseño del sistema de detección y alarma, y una revisión hidráulica del sistema de rociadores.

La evaluación de detección mostro que el sistema se encuentra adecuadamente implementado, detectores de calor lineales se encuentran distribuidos uniformemente en el área protegida permitiendo una detección adecuada del caso de diseño. Sin embargo, el cálculo hidráulico del sistema mostró deficiencias descargando la solución espumosa a una tasa menor al requerido por el estándar de diseño de sistemas de rociadores con espuma NFPA 16 [13].

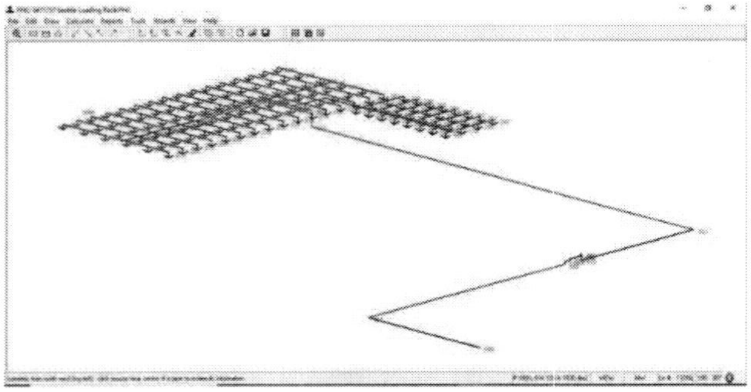
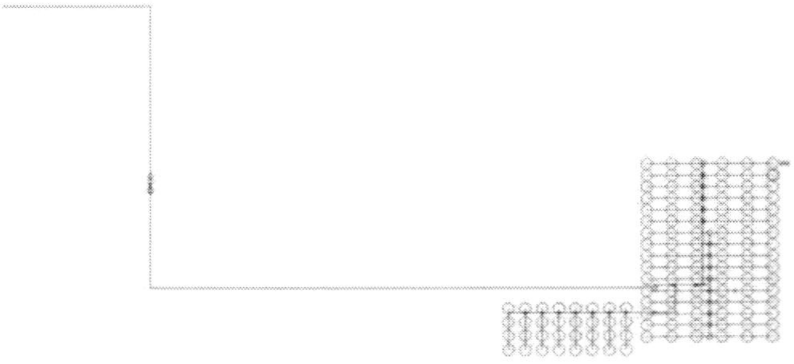

Figura 15. Cálculo hidráulico del sistema

3.5 *Taller FHA*

Luego de la evaluación de desempeño y el cálculo de consecuencias, se procedió a realizar el taller FHA donde se analizó toda la data producido por los cálculos de consecuencia y de desempeño.

3.5.1 <u>Descripción de escenario</u>

El taller empezó por la descripción del escenario de diseño. El escenario de incendio de diseño fue mostrad de la siguiente forma:

Ruptura total de manguera de carga durante el llenado del camión cisterna de gasolina. La ruptura ocurre en la conexión al camión produciendo un incendio tipo piscina/charco. El escenario de diseño considera un viento calmado ya que el incendio se produciría en un área cogestionada que disminuiría la acción del viento.

3.5.2 Filosofía de respuesta

La filosofía de diseño incluye la operación actual, la cual fue definida de la siguiente forma:

El sistema de protección contra incendio se activa automáticamente por acción del sistema de detección. Al detectarse un incendio, el sistema abre la válvula del agua descargando el agua por los rociadores. La activación de la espuma debe realizarse manualmente, el operador en el área de carga debe acercarse al cuarto de espuma y abrir la válvula de paso, la cual pasa concentrado de espuma al inductor de forma que la solución espumosa de 3% AFFF sea descargada por los rociadores.

Se realizó una evaluación del sistema de rociadores de espuma donde el cálculo hidráulico del sistema indicó que no cumple con las especificaciones de la NFPA 16. El sistema no incluye bomba contra incendios o estanque de almacenamiento y depende completamente del flujo de agua municipal. Al tener el sistema tipo diluvio completo, el cual incluye todos los carriles la presión y flujo en los rociadores no es la requerida por NFPA 16.

3.5.3 Detección de incendios

La detección contra incendio es realizada por el sistema de detección automática instalado. Este sistema consiste en un detector lineal de calor, conectada a una central de incendio. Cuando el detector lineal experimenta una temperatura de 70°C la central de incendio lo registra como un evento.

Debido a que el sistema es automático, se consideró un tiempo máximo de 2 minutos de respuesta de agua. Sin embargo, la introducción de espuma seria mayor ya que el operador debe acercarse al cuarto de espuma y abrir la válvula. El equipo del taller consideró un tiempo máximo de 5-10 minutos para abrir. Sin embargo, se consideró que estos tiempos deberían ser revisados una vez se presente los resultados del modelo de consecuencias.

3.5.4 Revisión de modelos

Se presentaron los cálculos de consecuencias enfocándose en el escenario de ignición temprana. Se consideró que el escenario de ignición tardía podría ser complejo ya que puede producir una explosión lo cual estaría fuera de alcance del FHA. Se presentaron los resultados de los modelos empíricos y el caso de CFD.

Se mostraron los perfiles de temperatura y flujos de calor en las paredes del modelo CFD.

3.5.5 Análisis de impacto

Al revisar el modelo CFD el equipo consideró que hay una posibilidad que el operador sea impactado durante la ignición y sea inhabilitado a abrir la válvula de espuma. Por otro lado, la descarga de agua no sería efectiva ya que no cumple con los criterios de NFPA 15 [14]. Por lo contrario, si hubiera una descarga automática de espuma el efecto de las llamas podría ser controlado aun cuando el incendio afectara al operador.

Al revisar las temperaturas en las paredes, se estimó que el incendio podría escalar si otra hubiera otra cisterna en el carril adyacente. De igual forma, la duración del incendio es considerado suficiente para colapsar parte del techo. Se consideró que dicho escenario podría tener un impacto significativo a la continuidad operativa del terminal.

3.5.6 Filosofía de protección

Durante el taller se consideró que el escenario de diseño no podía atacarse de forma manual y que se debía mejorar el sistema de protección instalado para que sea capaz de controlar dicho incendio automáticamente. Revisando los resultados del cálculo consecuencias se pudo entender que había una posibilidad de que el personal en el área de carga fuera impactado por el evento imposibilitando su acción de emergencia. De tal forma, que se decidió que el sistema instalado debía ser automático con posibilidad de activación dese del cuarto de control de ser necesario. De la misma forma se decidió que el sistema de detección debe enviar una señal de alarma al cuarto de control y no solo una alarma general.

Al revisar la evaluación de desempeño se pudo discutir las posibles mejoras al sistema para asegurar su confiabilidad, evitando activaciones accidentales o fallas en el sistema durante su operación. Uno de los problemas principales de desempeño observado es que el sistema estaba diseñado para descargar el agente extinguidor en toda el área de operación, mientras el cálculo de consecuencia indica que la llama abarca inicialmente un área menor. De tal forma, que se discutió la posibilidad de limitar la activación a 3 carriles, disminuyendo la descarga en áreas no afectadas generando un aumento de flujo y presión en las áreas afectadas; vale acotar que el área de tres carriles correspondía al área mínima de diseño recomendada por la NFPA 16 [13].

3.5.7 Mejoras y recomendaciones

El taller FHA logró identificar una filosofía de diseño óptima y recomendó la evaluación de cálculos hidráulicos considerando la activación de solo tres carriles, y la iniciación de un proyecto de mejora del sistema para asegurar la activación individual de cada carril con la filosofía de la activación de 3 carriles. De la misma manera, recomendó la realización de un análisis de confiabilidad del sistema de detección y alarma para reducir la posibilidad de la activación accidental.

Al finalizar el estudio FHA, se logró identificar el verdadero problema de protección, evaluar diversas opciones para controlar adecuadamente el riesgo presente y encontrar la manera óptima de controlar dicho riesgo de forma efectiva.

4 Conclusión

La protección contra incendio es un elemento fundamental en el control de riesgo, ya que establece medidas mitigativas que mitiga los efectos producidos por un incendio accidental. Esta mitigación ayuda a disminuir los efectos del evento accidental y mantener las operaciones de la organización.

De tal forma, que estos sistemas deben ser diseñados apropiadamente y el FHA permite la comparación del desempeño del sistema con el riesgo presente permitiendo un diseño adecuado y óptimo.

5 Referencias

[1] CCPS. *Guidelines for Fire Protection in Chemical and Hydrocarbon Processing Facilities.* Center for Chemical Process Safety, American Institute of Chemical Engineers, first edition, New York, NY, 2003

[2] CCPS. *Guidelines for Hazard Evaluation Procedure.* Center for Chemical Process Safety, American Institute of Chemical Engineers, second edition, New York, NY, 1992.

[3] ISO. ISO17776 Petroleum and natural gas industries – Offshore production installations – *Guidelines on tools and techniques for hazard identification and risk assessment.* International Standards Organization, Geneva, Switzerland, 2000.

[4] Dow. Dow's Fire & Explosion Index Hazard Classification Guide, American Institute of Chemical Engineers, Seventh Edition, 1994

[5] SFPE. The SFPE Handbook of Fire Protection Engineering, Section 5 Fire Risk Indexing (J. M. Watts, Jr), p. 5-170. National Fire Protection Association, Fourth edition, Quincy, Massachusetts, 2008

[6] SFPE, Fire Risk Assessment, Engineering Guide SFPE G.04 2006, SFPE, Maryland, 2016

[7] SFPE, Guidelines for Substantiating a Fire Model for a Given Application, Engineering Guide SFPE G.06 2011, SFPE, Maryland, 2011

[8] CCPS. *Guidelines for Consequence Analysis of Chemical Releases.* Center for Chemical Process Safety, American Institute of Chemical Engineers, first edition, New York, NY, 1992

[9] CCPS. *Guidelines for Fire Protection in Chemical and Hydrocarbon Processing Facilities.* Center for Chemical Process Safety, American Institute of Chemical Engineers, first edition, New York, NY, 2003

[10] CCPS. *Guidelines for Risk Based Process Safety.* Center for Chemical Process Safety, American Institute of Chemical Engineers, New York, NY, 2007.

[11] Ministerie van Verkeer en Waterstaat, Publication Series on Dangerous Substance (PGS 3), Guideline for quantitative risk assessment, "Purple Book," CPR 18E, 2005

[12] CCPS. *Guidelines for Determining the Probability of Ignition of a Released Flammable Mass.* Center for Chemical Process Safety, American Institute of Chemical Engineers, New York, NY, 2014.

[13] National Fire Protection Association (NFPA), NFPA 16: Standard for the Installation of Foam-Water Sprinkler and Foam-Water Spray Systems, NFPA, 2019 edition

[14] National Fire Protection Association (NFPA), NFPA 15: Standard for Water Spray System for Fire Protection, NFPA, 2017 edition

618848 Investigation of Thermal Hazards in Reaction Systems

Abstract:

Effective process safety programs provide for the prevention and control of hazards, which result in risk reduction and thus sustained value. A major source of hazards within the chemical process industries involves process scale up, as risks of thermal runaway and explosions are introduced. Thermal runaway occurs when the heat produced in a chemical reaction exceeds the heat removed, which results in an uncontrolled accelerating rate of reaction. An example of a process scale-up incident is the MFG Chemicals facility in Dalton, GA. For this incident, a reactor containing 4000 gallons of triallyl cyanurate overheated, resulting in an explosion that caused 154 injuries. Scale-up incidents such as these often result from a lack of understanding of the reactive chemistry hazards associated with the reaction being carried out as well as an ineffective process design and hazard review, which can result in insufficient heat removal capabilities.

In this research, the Mettler Toledo RC1 Calorimeter, an isothermal calorimeter, is used to investigate thermal hazards associated with reaction systems found within the process industries, particularly oxidation reactions. This provides useful data at the lab scale that can be used for safe process scale up as well as analysis of hazards to identify necessary layers of protection to be implemented. As oxidation reactions tend to be highly exothermic, thorough understanding of the reactive chemistry mechanisms is critical for scaling up the reactions for use in industrial settings.

For this work, the oxidation reactions of 2-butanol to 2-butanone and 2-pentanol to 2-pentanone, which are catalyzed by titanium silicalite-1, are investigated. For these, the reaction temperature, reactant quantity, and catalyst quantity are varied. The effects of these changes on the temperature profiles, heat release rates, and maximum adiabatic temperature increases are investigated. Based on these results, potential hazards are identified and methods for safely scaling up these reactions are proposed.

Seven Reasons to Pilot a Chemical Process

Jeff Hippler

Mid-Atlantic Technology, Research, & Innovation Center

MATRIC

Pilot plants are expensive to build and operate. Many companies are finding themselves without the facilities or expertise to build a full pilot plant that can be scaled to a commercial venture. The good news is that the need to build these pilot plants is not as critical as it once was. However, despite our industry's ability to model and simulate practically every aspect of our chemical processes, there is still often a need to build a pilot unit.

Pilot units are built to de-risk a commercial venture. They can be thought of as insurance policies to identify and fix issues before they crop-up at a more expensive scale. Occasionally, these pilot units are built concurrently with the commercial unit to speed the time to market while also providing a troubleshooting tool to minimize any downside risk.

When contemplating whether to go with a pilot plant, it is important to delineate exactly why a pilot plant should be built. Fortunately, we are well past the days when they are built because 'we have always done it that way'. But it is still an invaluable exercise to describe exactly when and why to pilot a process and define the ultimate measure of success.

This talk covers seven reasons to build a pilot plant for an emerging process and includes examples of each of these situations to consolidate the thought processes around this critical step during process commercialization

1. Solids
2. Recycle
3. Impurities / Specifications
4. Catalyst stability / Lifetime
5. Material of construction / Process Equipment
6. Phase behavior
7. Process samples

Using this list as a reference to frame any discussions about what to pilot and the scale to pilot can generate agreement among a team more quickly and help in working toward your best solution.

619147 Scalable Pervaporation Membrane to Enable New Continuous Flow Chemical Synthesis Routes

Abstract:

Pervaporation membranes can be used to avoid azeotropic distillation, enable flow chemistry technology, degas organic solvents, dehydrate solvents, drive reactions towards product formation and more. Compact Membrane Systems (CMS) has demonstrated the effectiveness of their fluoropolymer pervaporation membranes to degas and dewater organic solvents for the pharmaceutical and specialty chemical industries in recent years. This has particular relevance in flow or continuous chemistry, which has gained momentum as the world moves to energy efficient, modular manufacturing.

A major challenge for the adoption of flow chemistry is transitioning from the laboratory bench to commercial scale. Teflon AF 2400 extruded tubing is the embodiment of this challenge. Researchers have successfully demonstrated the feasibility of flow chemistry reaction mechanisms at the benchtop scale, but many of these chemistries have not been implemented at large scale[1,2]. This talk will present a side by side by side comparison of a CMS pervaporation membranes to Teflon AF tubing as a scalable, cost effective alternative solution. Oxygen removal rates were studied at ambient temperature at a variety of flow rates and concentration scenarios. The CMS membrane was able to match or exceed the Teflon AF 2400 tubing in recovery and degassing across all cases studied. Economic models also show that the CMS membrane is both an efficient and cost-effective solution. A case study will also be presented to show the effectiveness of the solution in the pharmaceutical and fragrance industries.

This case study was performed in collaboration with Snapdragon Chemistry and the Massachusetts Institute of Technology and published in OPRD in 2020[3]. A CMS membrane was used to selectively remove ethylene from a ring closing metathesis reaction to drive product formation and enable the use of flow chemistry.Using traditional batch techniques, implementing olefin metathesis reactions on the commercial scale has proven challenging, if not impossible. Ring closing reaction mechanisms are commonly used during the preparation of small molecules for the pharmaceutical and fragrance industries. Membranes are a modular, scalable solution that can continuously remove the ethylene byproduct and enable the use of commercial scale continuous ring closing metathesis. This case study details the advantage of using a sheet-in-frame membrane reactor as opposed to a stainless steel tubular reactor. The fluoropolymer membrane has excellent chemical and thermal stability which allows for an extended range of thermal conditions without sacrificing selectivity or flux. Pure ethylene flux of the membrane was also investigated to ensure that the mass transport across the membrane would always be greater than the reaction's ethylene production rate. Scale up considerations including residence time, required ethylene removal rate and module design will be addressed. The feasibility of using a membrane reactor to remove byproducts shown in this case study is only one example of the reactions that could benefit from the implementation of membrane reactors.

References:

(1) Hughes, D.; Wheeler, P.; Ene, D. Olefin metathesis in drug discovery and development - examples from recent patent literature.

Org. Process Res. Dev. 2017, 21, 1938-1962.

(2) Higman, C. S.; Lummiss, J. A. M.; Fogg, D. E. Olefin metathesis

at the dawn of implementation in pharmaceutical and specialty chemicals

manufacturing. Angew. Chem., Int. Ed. 2016, 55, 3552-3565.

(3) Breen, C.; Parrish, C; Shangguan, N.; Majumdar, S.; Murnen, H.; Jaimson, T.; Bio, M. A Scalable Membrane Pervaporation Approach for Continuous Flow Olefin Metathesis. Org. Process Res. Dev. 2020, 24, 2298-2303

In 2018, Mike Roman (3M Chairman and CEO) established the 4th pillar of the 3M Strategy, titled "People and Culture" with the strategic goal of strengthening our competitive edge in attracting top talent as well as developing our people & advancing our culture. The three prongs to this pillar include: inclusive workplace, development for all, and future of work (flexibility/global/digital). To continue to build on the strength of 3M's inclusive culture, the CEO Inclusion Council (headed by Mike) was established which includes the Employee Resource Networks' leaders along with Inclusion Leaders in division & staff groups. I was named the Inclusion Leader for the Corporate Research Division in late 2018 and will discuss a variety of Case Studies (best practices and key learnings) from my experience over the last two years.

In my talk, I will cover three different areas: psychological safety, personal advocacy, and organizational leadership. I will discuss my own journey from awareness to allyship to advocacy. I will present case studies including what you can do: mentorship, speaking out for yourself and others, moving beyond your company and into the community as well as driving metrics that facilitate change. Finally, my presentation is meant to help others create an environment where people can show up as their authentic selves and flourish in a diverse, inclusive and equitable environment.

621648 Case Study: EDI in Practice

Abstract:

Part 2 of the session is a case study from a plant manager of DuPont, Jana Green, who took action with her employees after the George Floyd murder, to foster small group discussions. Her advanced EDI practices illustrate how she created psychological safety for the employees to share their feelings, ways she facilitated the group discussions with an approach that was mindful of different stakeholder groups. She exhibited high emotional intelligence throughout the discussions and went on to share them operations leaders across the U.S.

619471 2021 AIChE Spring Meeting & 17th GCPS Wednesday Keynote: Advancing the Circular Economy

Abstract:

The past several decades have illustrated the integral role plastics play in our society. From enabling modern conveniences to playing a role in the advancement of our sustainability goals, plastics are key.

Looking to the future, LyondellBasell is taking a leading role in advancing the circular economy. Ken Lane, LyondellBasell's executive vice president of Global Olefins & Polyolefins, will discuss the company's ambitions and comprehensive efforts to eliminate plastic waste in the environment.

621135 *Great Engineering Skills and High Emotional Intelligence: A Winning Combination*

Abstract:

Every one of us in the field of chemical engineering has invested many hours in the learning and application of principles of *chemistry*, biology, physics, and math to solve problems that involve the production or use of *chemicals*, fuel, drugs, food, and many other products. Indeed, chemical engineers apply mathematics, chemistry and other natural sciences to develop practical ways of using energy and materials for the benefit of society. I love to say we make relevant and impactful chemistry and processes happen. But what about the soft skills? This is where Emotional Intelligence becomes a must have. In this presentation, we will highlight how emotional intelligence or the ability to identify, understand and manage emotions is critical in STEM fields and especially during times of crisis. We will explain how self-awareness and self-regulation, intrinsic motivation, empathy and social skills can help you go further in your career and are must haves for managers and leaders of today.

618409 Enhancing Leadership and Management of Individuals and Teams through Emotional Intelligence

Abstract:

Academic performance and job opportunities in engineering and science positions typically place a premium on intelligence (IQ) and analytical ability. Although these are critical competencies for successful employees and leaders/managers, they are often not the determining factor for success in technical positions. Leaders and managers must be able to understand, interpret, and use their and others' emotions to attain goals and function effectively personally and on teams; this requires emotional intelligence (EI or EQ). Traditional engineering or science programs do not address or even mention EI, although the lack of this skill set is evident during specific course requirements and in various career assignments: whenever teams must function efficiently and effectively. The absence of this trait generally leads to frustration and anxiety for leaders and followers and inhibits many competent and motivated engineers and scientists from pursuing or remaining in leadership/management positions. This talk will describe characteristics of technically-trained individuals who lack EI and the effect that this deficiency has on the individual, on team effectiveness and on the attitudes of subordinates, peers, and superiors. Approaches to improve or develop EI and thereby enhance leadership/management prowess will be discussed.

SPRING21
+17TH GCPS
A Joint AIChE and CCPS Meeting

Slaying the Dragons that Sabotage Protection Measures to Feed on the Ash of Your Plant

Denise Chastain-Knight
Exida Engineering, LLC
Sellersville, PA
dchastainknight@exida.com

Prepared for Presentation at
American Institute of Chemical Engineers
2021 Spring Meeting and 17th Global Congress on Process Safety
Virtual
April 18 - 22, 2021

Slaying the Dragons that Sabotage Protection Measures to Feed on the Ash of Your Plant

Denise Chastain-Knight
Exida Engineering, LLC
Sellersville, PA
dchastainknight@exida.com

Keywords: Hazard Identification and Risk Analysis, Safety Instrumentation Systems, Measurement and Metrics

Abstract

Dragons are legendary winged monsters that breathe fire and feed on ash. Or so we think. This is a tale of three dragons that have evolved to hide in plain sight and quietly feed on the underpinnings of protection layers, eroding the effectiveness of process safety mitigation measures. These dragons are mischievous and feed off process upsets, near misses and accidents. The Invisible dragon roams around the plant breaking devices and causing people to make mistakes that set off alarms and causes Safety Instrumented Systems (SIS) trips to initiate. Its special joy is creating initiating events that were not considered in design. Tiny and Giant dragons specialize in defeating the SIS protection layers, so that all three dragons can feed on the ash of an accident. The Tiny dragon baits SIS designers with devices having extremely low failure rate data causing the humans to believe Safety Instrumented Functions (SIF) are more reliable than they actually are. The Giant dragon works alongside the Tiny dragon to further undermine the SIS with its song of near perfect proof testing. This paper will illustrate the impacts of underestimating SIF demand rate, using overly optimistic failure rate data and overestimating proof test coverage. We will discuss techniques to identify and correct these issues and suggest metrics to monitor SIS performance thus slaying the dragons.

1 Introduction

Safety Instrumented Functions (SIF) are intended to mitigate a specific amount of risk based on the potential consequences of a specific hazard scenario as determined by the hazards risk assessment. The design of the SIF involves careful selection of components and operating and maintenance parameters, in order to deliver the required risk reduction. Verification and validation of the SIF performance is measured against the stated design specification. Failure to maintain any of the specification parameters can result in SIF performance that does not meet the design intent.

The risk reduction requirements for a SIF, in low demand, are stated in terms of Safety Integrity Level (SIL) target and a specific Risk Reduction Factor (RRF). For example, a low demand SIF with a target SIL 2 RRF of 225 must achieve an average probability of failure on demand (PFD_{avg}) of 4.4E-3 and meet SIL 2 hardware fault tolerance and systematic capability. Hardware fault tolerance is the physical redundancy as required in IEC 61511[1] or IEC 61508 [2] standards to address random hardware failures. Systematic capability is essentially the systematic fault tolerance and is assessed through certification or Prior Use justification. The discussion in this paper addresses factors that primarily impact the PFD_{avg} ($PFD_{avg} = 1/RRF$). The three dragons represent input parameters of SIL verification calculations where PFD_{avg} is assessed. This paper will illustrate how PFD_{avg} is influenced by demand rate, component failure rate and proof test coverage. It will also discuss how failure to maintain operation consistent with these parameters can result in SIFs that do not achieve the intended risk reduction target.

2 Invisible Dragon – System Demands

Design and verification rules for a SIF are determined based on the mode of operation. SIFs may operate in Low Demand, High Demand or Continuous mode. Low Demand mode is quite common in the process industry and random failure rate is assessed by PFD_{avg}. High demand and Continuous mode failure rate is defined as probability of failure per hour, or PFH. For a SIF to be operating in Low Demand mode, the initiating event frequency must be less than 1/(2*PTI)[3] where PTI is the proof test interval. A SIF with an initiating event frequency that exceeds the Low Demand threshold is considered High Demand and must be designed per the High Demand mode criteria, thus there is no benefit derived from proof testing. The effect of proof test interval on SIF demand mode is illustrated in Figure 1.

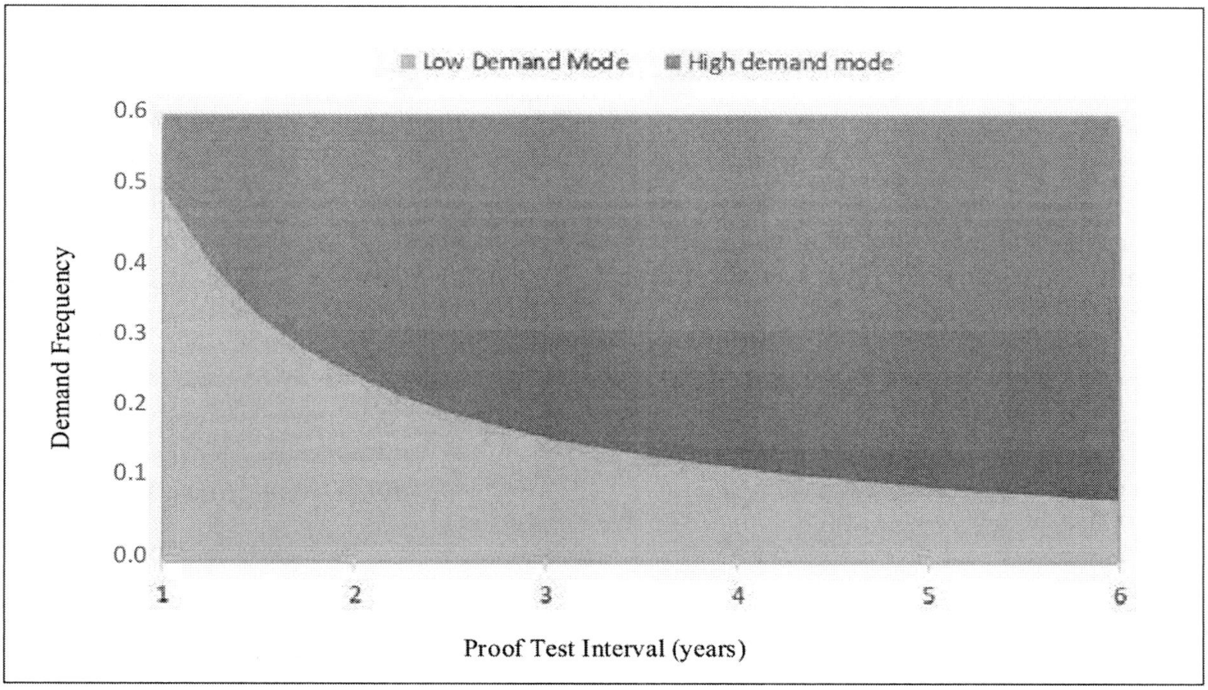

Figure 1 - SIF Demand Mode Threshold

Consider the process illustrated in Figure 2. During the LOPA, a scenario was developed resulting in a SIF to protect against low flow from V-100 to V-200.

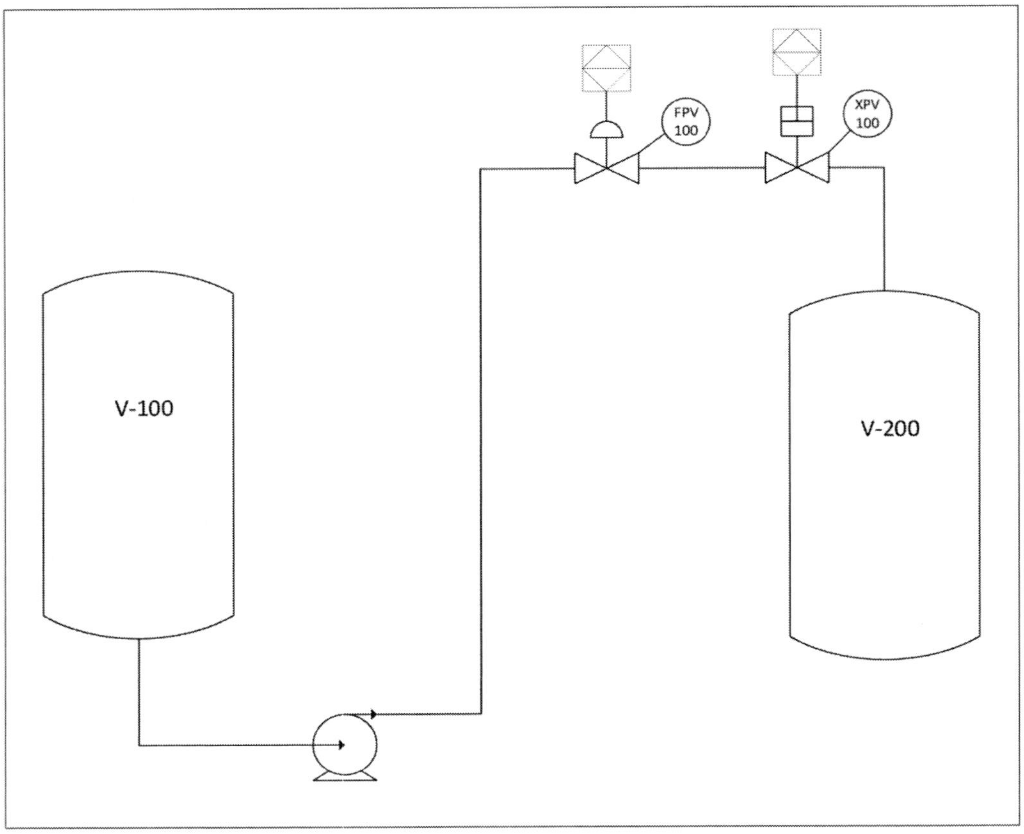

Figure 2 – Example Process

The LOPA given in Figure 3 considers pump failure, FPV failure and XPV failure, as initiating events for the SIF. Demand on the SIF is reduced for the pump failure case by a fault alarm with operator action. The SIF demand was determined to be 0.21/year. Proof test interval for the SIF is 2 years. The SIF demand threshold is calculated to be 1/(2*2) or 0.25/year so the SIF is operating in low demand mode.

The original design for the SIF is very near the threshold and could easily be pushed into high demand by an increase in proof test interval or the initiating event frequency. For example, if the facility decided to delay a turnaround by 6 months, the proof test interval is then increased to 2.5 years changing the high demand threshold to 1/(2*2.5) or 0.2/year. With this change, the SIF is now effectively operating in high demand mode.

1. Low of flow from V-100 to V-200 / Initiating Event	Frequency [per year]	IPLs: Pumpt Fault alarm		IPLs: Loss of flow SIF		Intermediate Frequency [per year]		Comments
Pump failure	0.1	B	0.1	B	0.01	B	1.00E-4	
		E	0.1	E	0.01	E	1.00E-4	
		S	0.1	S	0.01	S	1.00E-4	
FPV-100 failure	0.1	B	NA	B	0.01	B	1.00E-3	
		E	NA	E	0.01	E	1.00E-3	
		S	NA	S	0.01	S	1.00E-3	
XPV-100 failure	0.1	B	NA	B	0.01	B	1.00E-3	
		E	NA	E	0.01	E	1.00E-3	
		S	NA	S	0.01	S	1.00E-3	

Figure 3 – Example Process LOPA

Because a decision to delay a turnaround is a conscious choice, an engineer can pause to perform SIL verification using high demand criteria quantifying the risk that is assumed during the delay. They can then implement additional protection and/or mitigation measures to mitigate that risk as needed.

But what if the initiating event frequency is higher than estimated in the LOPA? This is the Invisible Dragon causing additional risk to be accepted without the benefit of foreknowledge, so mitigation measures are not considered. For example, presume the FCV sticks periodically causing low flow and operators are accustomed to unsticking the valve when needed. If the actual failure frequency for the FCV is slightly higher, say 0.15 failures/year, than the assumed literature value (0.1 failures/year), the overall demand rate is actually 0.26/year and the SIF is always operating in High Demand mode.

How can the Invisible Dragon be made visible? Consider implementing metrics to monitor SIF demand rate or use forensic analysis to confirm that it is within the design basis. If the SIF demand is found to be borderline, similar to the example process, consider implementation of additional independent protection layers (IPLs) to reduce the demand so that the SIF remains in Low Demand mode at all times.

3 Tiny Dragon – Unrealistic Failure Rate Data

The SIL verification calculation equation for PFD_{avg} can become very complex as terms are included to address factors such as proof test effectiveness and diagnostics. The dangerous failure rate λ_D, whether detected (λ_{DD}) or undetected(λ_{DU}), appears in every term. If the verification calculation is to be realistic, then it is critical that the underlying failure rate data be realistic. The Tiny dragon baits SIS designers with devices reporting extremely low failure rate data causing the people to believe SIFs are more reliable than they actually are.

Figure 4 illustrates the structure of a SIF having a SIL 2 RRF 100 target. The SIF is designed to stop catalyst feed to a reactor when high pressure is sensed in the reactor. The sensor group includes three transmitters with two out of three (2oo3) voting. The final element group includes two identical valve assemblies consisting of a solenoid, actuator, and valve. The valve assemblies are placed in series in the catalyst feed line, so flow is stopped if either valve is successfully closed (1oo2 voting). A generic SIL 2 certified PLC is the logic solver. On the bottom left are the verification results (RRF 121.4) for the model using device data that has been vetted and is considered realistic. On the right, are the results (RRF 177) when the solenoid, actuator and valve reliability data is replaced with data from a certificate that appears to be optimistically low. The difference in the overall RRF achieved is significant. The significance of the difference is most obvious when the final element part of the SIF is compared in detail. The PFD_{avg} given by the optimistic data (8.7E-5) is less than 4% of that given by the realistic data (2.69E-3). Notice also that the Mean Time to Fail Spurious (MTTFS) for the final element group increases significantly from 40.03 to 159.77 suggesting the final element group is more resistant to spurious trips than it actually is. Finally note that the architectural capability (AC), or hardware fault tolerance (HFT), determined by the optimistic data suggests the final element arrangement could be used up to SIL 4, where the realistic data limits use to SIL 3.

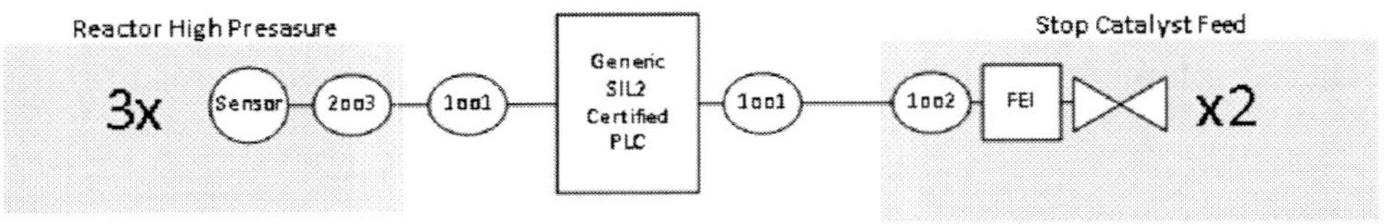

Vetted Valve Data

SAFETY INSTRUMENTED FUNCTION PERFORMANCE	
Target SIL	2
Target RRF	100
Achieved SIL	2
PFDavg	8.23E-03
SIL (PFDavg)	2
SIL (Arch. Constraints IEC 61508)	2
SIL (Systematic Capability)	N/A
Achieved RRF	121.4
MTTFS (years)	7.42

	PFDavg	MTTFS	SIL Limit	
			AC	CAP
Sensor Part	4.03E-03	549.13	2	N/A
Logic Solver Part	1.54E-03	9.26	2	N/A
Final Element Part	2.69E-03	40.03	3	N/A

Unrealistic Valve Data

SAFETY INSTRUMENTED FUNCTION PERFORMANCE	
Target SIL	2
Target RRF	100
Achieved SIL	2
PFDavg	5.65E-03
SIL (PFDavg)	2
SIL (Arch. Constraints IEC 61508)	2
SIL (Systematic Capability)	N/A
Achieved RRF	177
MTTFS (years)	8.62

	PFDavg	MTTFS	SIL Limit	
			AC	CAP
Sensor Part	4.03E-03	549.13	2	N/A
Logic Solver Part	1.54E-03	9.26	2	N/A
Final Element Part	8.70E-05	159.77	4	N/A

Note: Contribution of common cause failure is not included

Figure 4 – SIF Results Vetted vs Unrealistic Failure Rate Data

Clause 11.9.3 in IEC61511 states "the reliability data used when quantifying the effect of random failures shall be credible, traceable, documented and justified and shall be based on field feedback from similar devices used in similar operating environment.". Failure rate data is obtained from industry data sources, provided by the vendor with device certification documents, or collected from within the operating facility. Unfortunately, one does sometimes encounter a certificate or vendor literature that includes questionable failure rates. Figure 5 is a plot of failure rate data for solenoids, which includes data points from three industry databases, calibrated Failure Modes Effects and Diagnostics Analysis (FMEDA)[4] assessment results, and data reported on several certificates. While the industry data and the FMEDA are reasonably consistent, the four datapoints in red would appear to have overly optimistic data.

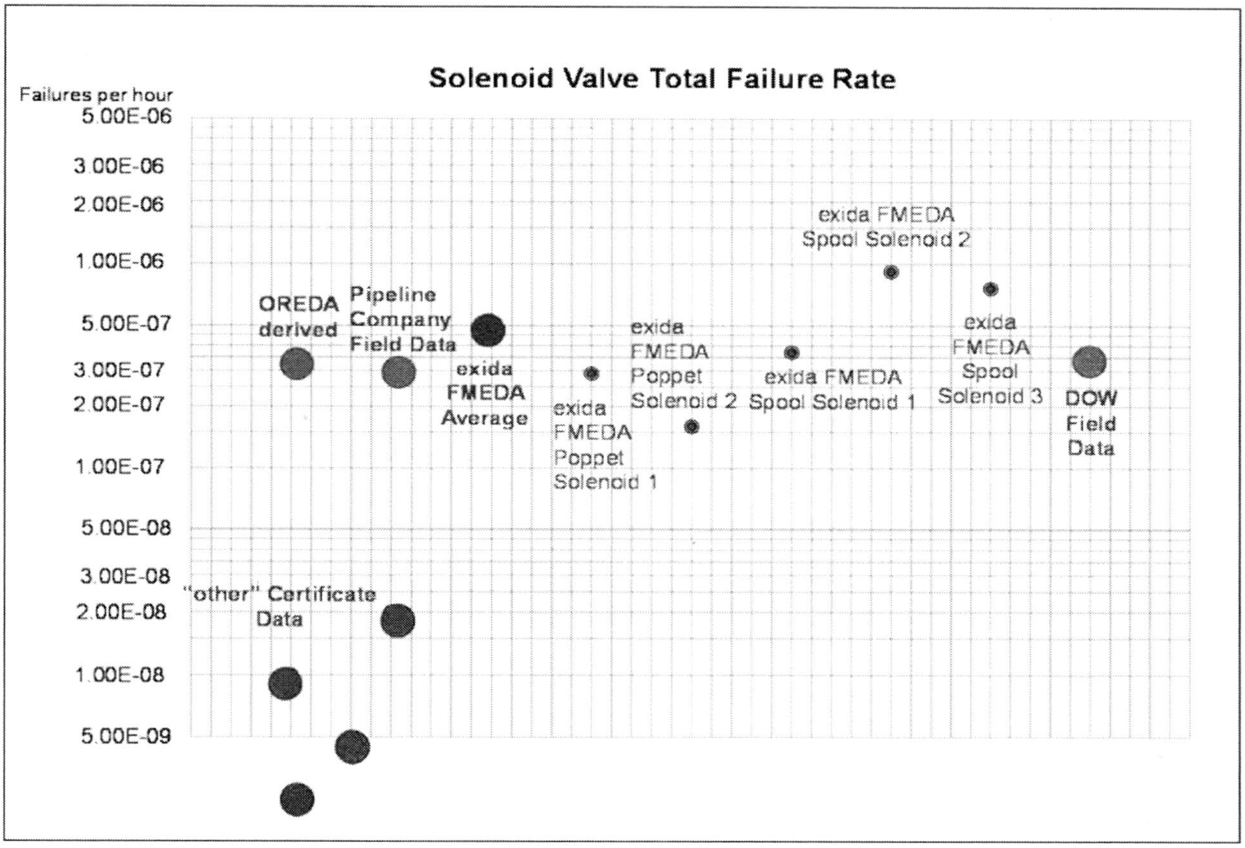

Figure 5 – Solenoid Valve Total Failure Rate

So how does one know if data presented by a vendor is realistic or not? The SILSafeData tool, http://silsafedata.com/, provides upper and lower boundaries for dangerous undetected failure rates for devices commonly used in process industries based on over 200 billion unit operating hours of field failure data. Designers should vet the failure rate information provided by vendors against industry data as part of the device selection criteria before selecting for use in a SIF.

4 Giant Dragon – Perfect Proof Testing

SIFs operating in low demand mode benefit from proof testing because there is opportunity to detect failures between demands and effect repairs. Proof test coverage (PTC) is the percentage of failures that can be identified with the proof test. While it is very tempting to assume that a proof test is perfect (100%), the reality is that proof tests are not always executed correctly or completely, or that the test may not effectively identify all failures[5]. The Giant Dragon undermines functional safety goals by using PTC values that are too large.

Figure 6 illustrates the impact of assuming perfect proof testing. The SIF is the same as described in Figure 4 above. The results shown on the left of the table include proof test coverage, which is based on careful analysis of the failure modes and the capability of the proof test to identify failures. The results (RRF 121.4) shown on the right are for the same SIF with the perfect proof test coverage in sensor, logic solver and final element groups.. The latter predicts a much higher overall RRF (181.3) and suggests the SIF is far more reliable that it actually is. The difference is less obvious in the final element group details, so one must review the PFD graph for the final element group shown in Figure 7 to appreciate that the higher PFD_{avg} is achieved through larger 'recovery' at proof testing.

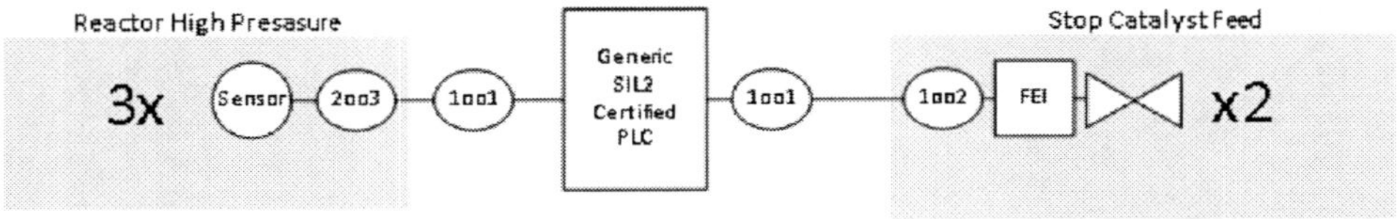

Realistic PTC Assumed

Safety Instrumented Function Performance	
Target SIL	2
Target RRF	100
Achieved SIL	2
PFDavg	8.23E-03
SIL (PFDavg)	2
SIL (Arch. Constraints IEC 61508)	2
SIL (Systematic Capability)	N/A
Achieved RRF	121.4
MTTFS (years)	7.42

	PFDavg	MTTFS	SIL Limit	
			AC	CAP
Sensor Part	4.03E-03	549.13	2	N/A
Logic Solver Part	1.54E-03	9.26	2	N/A
Final Element Part	2.69E-03	40.03	3	N/A

Perfect PTC Assumed

Safety Instrumented Function Performance	
Target SIL	2
Target RRF	100
Achieved SIL	2
PFDavg	5.51E-03
SIL (PFDavg)	2
SIL (Arch. Constraints IEC 61508)	2
SIL (Systematic Capability)	N/A
Achieved RRF	181.3
MTTFS (years)	7.41

	PFDavg	MTTFS	SIL Limit	
			AC	CAP
Sensor Part	2.64E-03	561.18	2	N/A
Logic Solver Part	1.54E-03	9.26	2	N/A
Final Element Part	1.35E-03	39.69	3	N/A

Note: Contribution of common cause failure is not included

Figure 6 – Imperfect Proof Testing Comparison

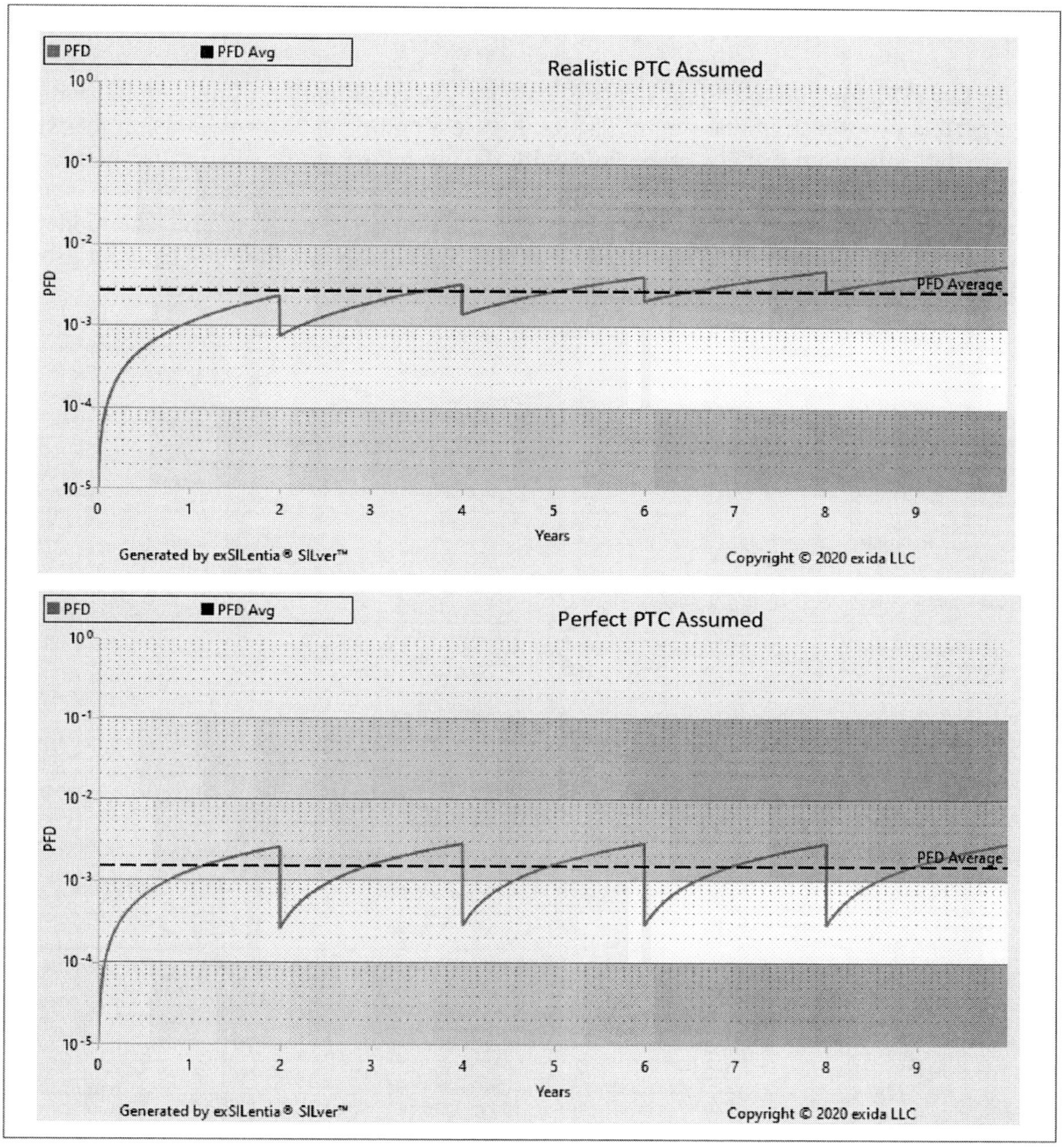

Figure 7 – Imperfect Proof Testing Comparison Final Element Group Graph

The analysis for the realistic model in Figure 6 and Figure 7 used PTC values from the software database that ranged from 51% to 99% for individual components. The proof test coverage of a device can be determined by FMEDA analysis[6] considering the proof test steps and the physical construction of the device. Many vendors will provide recommended proof test procedures with associated coverage in a product's safety manual.

5 The Tiny and Giant Dragons Team Up

When the Tiny Dragon and the Giant Dragon team up, the combination of unrealistic failure rate data and perfect proof testing compound results. Figure 8 compares the model using realistic failure rate data and proof test coverage (left) with the model using unrealistic final element data and perfect proof test coverage (right). The model with realistic failure rate and PTC basis delivers RRF 121.4 and the final element group is limited to SIL 3 architecture. The model using optimistic information delivers an overall RRF of 237.1 and suggests the final element group is acceptable for SIL 4 service.

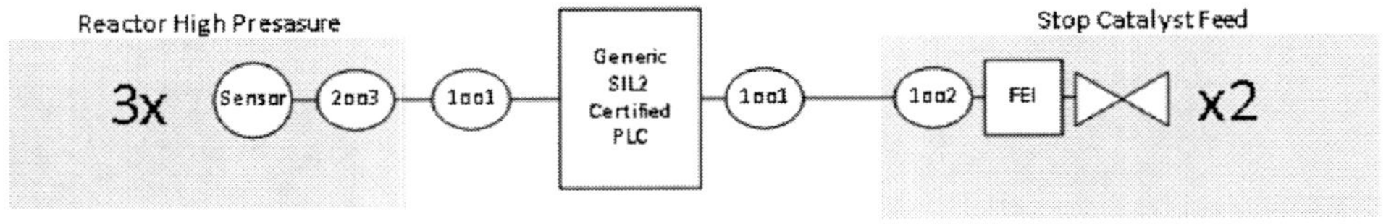

Realistic Failure Data and PTC

SAFETY INSTRUMENTED FUNCTION PERFORMANCE	
Target SIL	2
Target RRF	100
Achieved SIL	2
PFDavg	8.23E-03
SIL (PFDavg)	2
SIL (Arch. Constraints IEC 61508)	2
SIL (Systematic Capability)	N/A
Achieved RRF	121.4
MTTFS (years)	7.42

	PFDavg	MTTFS	SIL Limit	
			AC	CAP
Sensor Part	4.03E-03	549.13	2	N/A
Logic Solver Part	1.54E-03	9.26	2	N/A
Final Element Part	2.69E-03	40.03	3	N/A

Unrealistic Failure Data and Perfect PTC

SAFETY INSTRUMENTED FUNCTION PERFORMANCE	
Target SIL	2
Target RRF	100
Achieved SIL	2
PFDavg	4.22E-03
SIL (PFDavg)	2
SIL (Arch. Constraints IEC 61508)	2
SIL (Systematic Capability)	N/A
Achieved RRF	237.1
MTTFS (years)	8.62

	PFDavg	MTTFS	SIL Limit	
			AC	CAP
Sensor Part	2.64E-03	561.18	2	N/A
Logic Solver Part	1.54E-03	9.26	2	N/A
Final Element Part	4.83E-05	159.71	4	N/A

Note: Contribution of common cause failure is not included

Figure 8 – Unrealistic Failure Rate Data and Imperfect Proof Testing Comparison

Comparing the overall PFD charts in Figure 9 one can observe that the model using realistic PTC and failure rate information predicts periods where the actual RRF falls below 100 where the 'sawtooth' peaks cross the 10^{-2} line. With a mission time of 10 years, the chart shows that the SIF is achieving less than a RRF of 100, increasingly significant periods of time in years 6-10. To mitigate the risk, a designer might recommend that the mission time of the SIF be reduced and schedule refurbishment or replacement of components at year 6 instead of 10. The model utilizing unrealistic information does not indicate these increased periods of risk so the need for reduced mission time would go unrecognized.

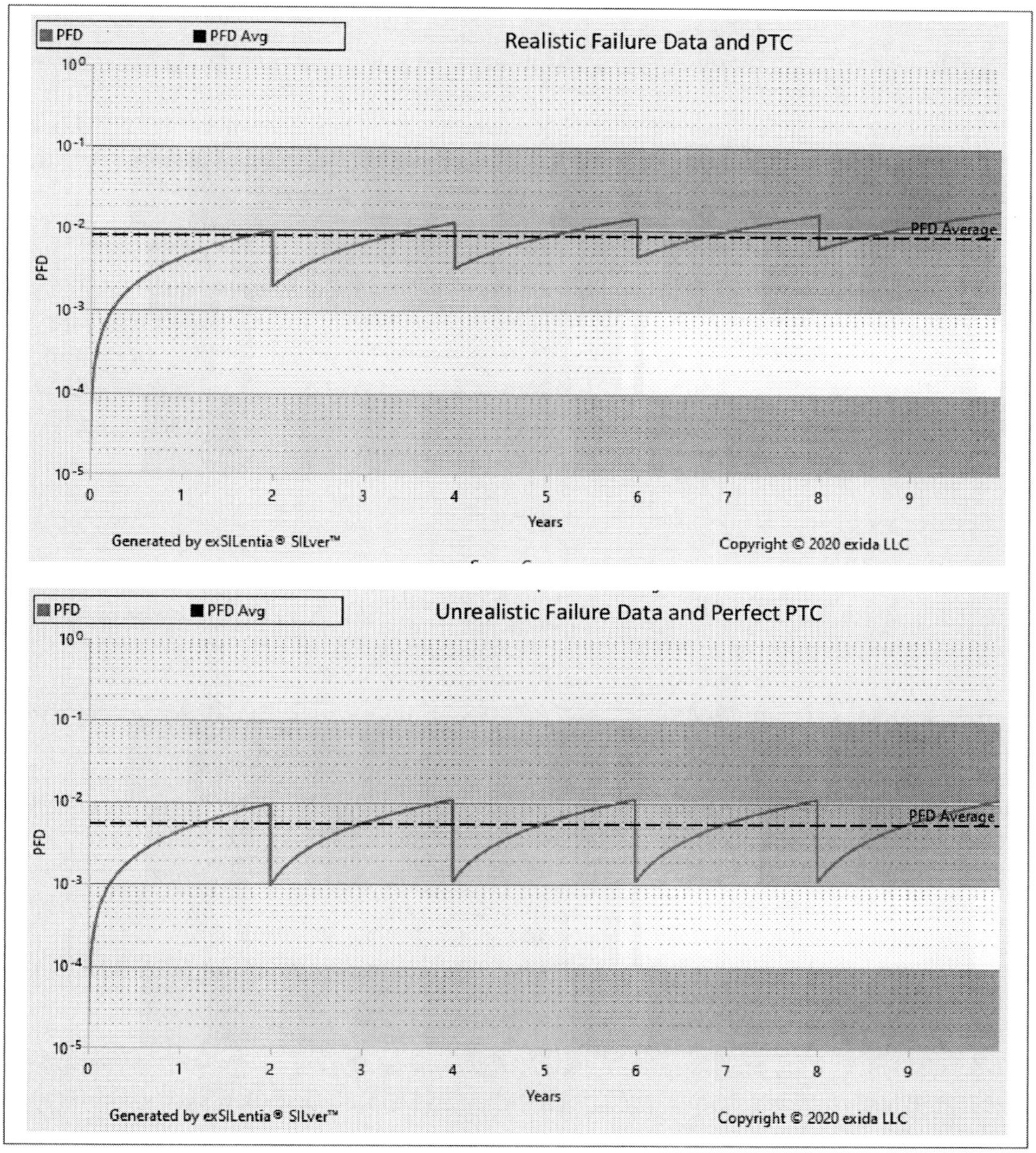

Figure 9 – Unrealistic Failure Rate Data and Imperfect Proof Testing Comparison Graph

6 Conclusion

What can designers do to slay the hungry dragons that sabotage protection layers, leading to the unknowing acceptance of additional risk?

Invisible: Remember that extending proof test interval of a SIF can move the operating mode from low to high demand. When adjusting PTI, designers should revisit SIL verification calculations in order to understand the risk that is assumed. Also, designers should periodically review the failure rate information for initiating events to confirm that the assumptions made during the PHA and LOPA remain valid.

Tiny and Giant: Don't fall for the lie of low failure rate data or the lure of near perfect proof testing. During device selection confirm that the data is credible, traceable, documented and justified per the requirements of IEC 61511 clause 11.9.3. Use of unrealistic failure rate data or assuming perfect proof test coverage can result in an overly optimistic PFD_{avg} calculation. When completing SIL verification calculations, avoid the use of overly optimistic failure rate information by vetting against industry data. Finally, use realistic proof test coverage for SIL verification based in the capability of the proof test procedure.

7 References

[1] IEC 61511-1 Ed 2.0, *Functional Safety: Safety instrumented systems for the process industry sector – Part 1: Framework, definitions, system, hardware and application programming requirements,* IEC, Geneva, Switzerland 2016

[2] IEC61508 Ed 2.0, *Functional safety of electrical/electronic/programmable electronic safety-related systems,* IEC, Geneva, Switzerland 2010

[3] Iwan van Buerden, William M. Goble, *Safety instrumented System Design, Techniques and Design Verification*, Instrument Society of America (ISA), Research Triangle Park, NC, 2018

[4] *Getting Realistic Failure Rate Data*, exida white paper, Sellersville PA, 2020. https://www.exida.com/Resources/Whitepapers/Getting-Realistic-Failure-Rate-Data (accessed March 1, 2021)

[5] Bukowski, J.V. and van Beurden Iwan, *Impact of Proof Test Effectiveness on Safety Instrumented System Performance*, Reliability and Maintainability Symposium (RAMS), Fort Worth, TX, January 2009.

[6] Goble, W. M., *Strategic Proof Testing*, exida white paper, Sellersville PA, 2020. https://www.exida.com/Resources/Whitepapers/strategic-proof-testing (accessed March 2, 2021)

"I Didn't Inhale, and Never Tried It Again."

Proper Handling of Toxic Inhalation Hazard (TIH) Chemicals in Plants and During Transportation

Michael P. Broadribb
Baker Engineering and Risk Consultants, Inc.
3330 Oakwell Court #100, San Antonio, TX 78218
mbroadribb@bakerrisk.com

Rafael Callejas-Tovar
Baker Engineering and Risk Consultants, Inc.
11011 Richmond Ave. #700, Houston, TX 77042
rcallejas@bakerrisk.com

Prepared for Presentation at
American Institute of Chemical Engineers
2020 Spring Meeting and 17th Global Congress on Process Safety
Virtual
April 18 - 22, 2021

"I Didn't Inhale, and Never Tried It Again."

Proper Handling of Toxic Inhalation Hazard (TIH) Chemicals in Plants and During Transportation

Michael P. Broadribb
Baker Engineering and Risk Consultants, Inc.
3330 Oakwell Court #100, San Antonio, TX 78218
mbroadribb@bakerrisk.com

Rafael Callejas-Tovar
Baker Engineering and Risk Consultants, Inc.
11011 Richmond Ave. #700, Houston, TX 77042
rcallejas@bakerrisk.com

Keywords: toxic hazards, process safety, refinery, chemical plant, transportation

Abstract

Numerous fatalities and serious health effects have occurred in the process industries through personnel exposure to toxic chemicals and materials. A Toxic Inhalation Hazard (TIH) chemical can be particularly hazardous as a release of some chemicals cannot be detected through sight or smell. TIH chemicals are widely handled in process units and also transported by road, rail, marine, and pipeline. Fatal incidents involving chlorine (Graniteville, SC, 2005), hydrofluoric acid (South Korea, 2012), and methyl mercaptan (MeSH) (La Porte, TX, 2014) (Ref.1) emphasize the risks of TIH chemicals if they are not handled with a sense of vulnerability and proper care. Most elements of CCPS Risk Based Process Safety (RBPS)(Ref. 2), plus some engineering design considerations, are important for the proper handling of TIH chemicals.

Transportation of TIH chemicals poses additional risks as an accidental release can potentially expose a large public population. In addition to other RBPS elements, hazard identification and risk analysis (HIRA) can be used to quantify the consequences and risks associated with transportation operations. A comprehensive quantitative risk analysis (QRA) involves the development of an event tree to identify scenarios, dispersion simulations to calculate vulnerabilities along the route, and a thorough frequency analysis to pair the scenarios with their likelihood to calculate risk. The results are typically presented in terms of FN curves[1], societal risk, and location specific individual risk, which are compared against risk tolerance thresholds to facilitate the decision-making process. These results have helped companies understand if the

[1] FN Curve is a plot of cumulative frequency (y-axis) versus consequences (x-axis), often expressed as number of fatalities. The value plotted on the y-axis is the cumulative frequency of experiencing N or more fatalities.

benefits of their transportation operations justify the associated risks, and to develop comprehensive risk mitigation strategies.

This paper describes the authors' experiences of applying: (i) good practices for handling TIH chemicals, (ii) a RBPS approach for assessing TIH risks at chemical plants, including loading/unloading (railcars, road tankers, ISO tanks), (iii) use of quantitative methods for TIH transportation risk decision-making, and (iv) real case examples of potential TIH risk reduction strategies.

1 Introduction

Toxic chemicals pose a potential risk if people are exposed to the chemicals, which includes not only the workforce within chemical plants and oil refineries, but also the local communities. The transportation of toxic chemicals by road, rail, marine, and pipeline also pose potential risks to the public and often travel through high density population areas. As a result, it is paramount that hazards are properly managed.

Toxic Inhalation Hazard (TIH) chemicals are a class of toxic chemicals that act on the respiratory tract and can lead to pulmonary injury. There are many industrial TIH chemicals that are widely handled in process units and transported, and some, such as chlorine bleach and ammonia, are commonly found in households. A TIH chemical can be particularly hazardous as a release of some chemicals cannot be detected through sight or smell. A fatal incident in South Korea that released a colorless volatile TIH chemical (hydrofluoric acid) is described in the case study below.

Hydrofluoric Acid Release, Gumi, South Korea, 2012 (Ref. 3)
Two workers at the Hube Global chemical plant climbed onto a road tanker to hook up an air hose intended to unload the hydrofluoric (HF) acid cargo by air pressure into a storage tank. While one employee connected the air hose, the other, presumably to save time, began unscrewing the bolts of the end plate on the acid outlet valve. He inadvertently bumped the manual outlet valve arm, which partially opened. When the valve opened, 8 tons of HF was released under pressure. During transportation, the cargo had become pressurized due to thermal expansion. Neither worker was wearing appropriate personal protective equipment (PPE) and were killed, as were three others on a neighboring facility. Another 18 workers were hospitalized, and 4,195 people, including 416 firefighters were treated for nausea, skin rash, and breathing difficulties. A significant area of arable land downwind of the release was contaminated, resulting in the loss of crops, over 3,900 cattle, and 1,200 corroded vehicles.
Lessons learned:

- Industry procedures were not followed. The tanker should have been de-pressured and appropriate flexible hoses deployed before efforts to commence unloading.
- Workers should wear proper acid suits with supplied air hoods.
- Municipal fire-fighters arrived without proper PPE and sprayed water on the release that intensified the rate of HF release.
- Local authorities were slow to evacuate the residents in the vicinity of the release, and deployed inadequate resources, such as neutralizing agents (lime).

2 Practices for Handling Toxic Chemicals

The first practice for handling toxic chemicals is to understand how to maintain primary containment of the chemical. One of the short definitions of process safety is:

Keep it in the Pipe!

This is especially true for TIH chemicals. It is essential to know if a chemical is corrosive, unstable, reactive with other chemicals, or has any other physical property that can impair primary containment. A Safety Data Sheet (SDS) should provide the necessary information. Obvious safeguards against a loss of primary containment (LOPC) include selection of appropriate materials of construction, avoiding inadvertent mixing with specific other chemicals, and appropriate storage and operating conditions. These and other safety measures are discussed below.

The next practice for handling toxic chemicals is to understand their physical properties in case of a LOPC, which should also be provided on the SDS. TIH chemicals can have acute or chronic health effects, so it is important to follow the required personal protection equipment (PPE) specified. Other key physical properties to understand are:

- If a release occurs, will it be a liquid or vapor/gas or aerosol?
  - Process conditions (temperature, pressure), ambient temperature, and if the TIH chemical is mixed with other chemicals, will determine which phase.
- If a liquid, know whether the TIH chemical is volatile, heavier or lighter than water, and if soluble.
- If a vapor or gas, know if the TIH chemical is heavier or lighter than air, and what happens when it cools.

When implementing capital projects and small management of change (MOC) projects that involve TIH chemicals, consideration should be given to inherently safer design (ISD) principles. These principles are:

- Intensification - minimize inventory of hazardous materials including TIH chemicals,
- Substitution - eliminate hazards by substituting less hazardous materials where possible,
- Attenuation - minimize operating severity by avoiding extremely high pressures, high temperatures, or low temperatures, and
- Simplification - avoid unnecessary complexity/redundancy to minimize potential release sources and facilitate operator understanding.

Other design considerations in addition to ISD involve risk management and reliability hierarchies, as detailed in Table 1 and Table 2 below.

Table 1. Risk Management Hierarchy

ELIMINATION is better than
PREVENTION is better than
CONTROL is better than
HAZARD MITIGATION is better than
EMERGENCY RESPONSE

Elimination is not always practical as it involves selection of alternative processes that do not handle TIH chemicals.

Some examples of PREVENTION include, but are not limited to:

- Corrosion – use of corrosion-resistant alloys,
- Joints – use of welded piping to prevent leaks/releases at flanges, screwed fittings and couplings,
- Rotating Machinery – use of canned pumps and double seals,
- Unmanned facility – automated remote operation to prevent injury/fatality,
- Secondary containment,
- Good operating, maintenance, and process safety practices (but may not be 100% successful – see reliability hierarchy in Table 2).

Examples of CONTROL to reduce size or frequency of the hazard include, but are not limited to:

- Emergency shutdown (ESD) systems – remotely operable, excess flow valves,
- Blowdown/dump systems – remotely operable,
- Process quench/neutralizing systems,
- Fire protection – deluge/sprinklers/monitors/foam - remotely operable,
- Restriction orifices,
- Minimize surface area of spilled liquid release,
- Manual intervention/shutdown – needs PPE,
- Good operating, maintenance and process safety practices.

Examples of MITIGATION of the impact on people include, but are not limited to:

- Secondary containment – dikes, water curtains,
- Toxic release general alarm – site and local community,
- Evacuation – escape packs / respirators, other PPE,
- Shelter-in-place – building tightness (doors/windows/louvres), pressurized (elevation of heating ventilation and air conditioning (HVAC) inlet), HVAC ESD (gas sensors, automation).

Table 2. Reliability Hierarchy

PASSIVE controls are more reliable than
ACTIVE controls are more reliable than
ADMINISTRATIVE or PROCEDURAL controls

Examples of PASSIVE controls include, but are not limited to:

- Containment dikes and kerbs,
- Restriction offices,
- Fire resistant cladding on structural steel,
- Smoke/fire dampers, fire walls, and fire doors,
- Blast resistant buildings.

Examples of ACTIVE controls include, but are not limited to:

- Emergency shutdown systems,
- Blowdown/dump systems,
- Process quench/neutralizing systems,
- Active fire protection – deluge/sprinklers/monitors/foam.

Administrative and procedural controls tend to be the least reliable as they involve human interaction, and human intervention in the case of an emergency or abnormal situation. Process units should have clear expectations for operator intervention in the event of a TIH emergency, who will require appropriate training and PPE. Similarly, if the facility has its own emergency response team or relies upon mutual aid and/or the local fire department/emergency medical technicians (EMT), they must have toxic release capability (knowledge and training/drills in headcount, personnel rescue, and stopping a release, plus all of the necessary equipment and PPE to do so).

3 Transportation of Toxic Inhalation Hazard Chemicals

Transportation of TIH chemicals is a key activity in the supply chain, which has been increasing due to market globalization. The transportation of TIH chemicals might involve long distances and different modes between the origin and destination, such as truck, rail, crane, ship, etc. Since the transportation operations of TIH normally use the existing infrastructure, the routes might pass through locations with high population, which poses additional risks as an accidental release can potentially expose a large public population. An incident in South Carolina released a volatile TIH chemical (chlorine) during transportation and impacted the local community resulting in fatalities, hospitalization, and evacuation. This event is described in the case study below.

Chlorine Release, Graniteville, South Carolina, 2005 (Ref. 4)

Many trains traveled through Graniteville each day. A rail spur in the center of the town serviced the Avondale plant, which received daily deliveries of chlorine by rail for mill operations. On January 5, 2005, the regularly assigned conductor and engineer were both off duty, and standby employees operated a local train that was parked in a siding near the Avondale plant. The switch connecting the spur to the main line had to be turned manually for deliveries. Two track warrants protecting the local train's use of the tracks were cleared. Although the railroad switch for the Avondale siding was supposed to be set for mainline operation before the train crew departed for the night, the train's brakeman later told the NTSB that he was "not 100 percent sure" he had aligned the switch for mainline operation. No train operated through the area or used the affected tracks for the rest of the day.

At 2:38 am on January 6, 2005, a freight train approached Graniteville at 48 miles per hour. When the engineer saw the improperly aligned switch, he emergency braked, but there was insufficient distance to stop. The freight train was diverted by the improperly lined switch onto the siding and collided with the local train. The collision derailed 3 engines and 18 freight cars.

The freight train was transporting chlorine, sodium hydroxide, and cresol. A railcar loaded with 90 tons of chlorine ruptured and the liquefied chlorine rapidly vaporized. Nine people died of chlorine inhalation (eight at the time of the accident, one later), and at least 554 people were treated for chlorine exposure. About 30% of the chlorine was recovered by industrial responders. While the fire chief ordered shelter in place and initiated a major evacuation, these actions did not take place immediately. In total, 5,400 residents within one mile of the crash site were forced to evacuate for nearly two weeks while hazardous materials (HAZMAT) crews decontaminated the area.

Lessons learned:

- The incident could have been avoided if railroad employees had been alerted to the switch error by an electronic signal.
- The Federal Railroad Administration (FRA) sets guidelines and standards for railroad companies and lines. However, the railroad company, main line, and rail siding located in Graniteville complied with these standards that were inadequate in 2005.
- Regular shipments of hazardous materials through populated areas potentially expose the public to significant hazard.
- The incident illustrated the difficulties of achieving adequate and efficient evacuation for hazardous materials events. The local community did not understand how to react (shelter in place, evacuate) to a chlorine release.

In addition to other RBPS elements, hazard identification and risk analysis (HIRA) can be used to quantify the consequences and risks associated with transportation operations. A comprehensive quantitative risk analysis (QRA) can provide the risk profile along the length of the route and help to identify potential high-risk locations and operations. This powerful insight can be used to develop cost-effective risk mitigation, if needed, and facilitate the decision-making process.

The main steps involved in the QRA are:

1. Identification of the scenarios to be considered.
2. Consequence analysis (dispersion and vulnerability calculations).
3. Frequency analysis.
4. Risk calculations.
5. Comparison of results vs. risk tolerance thresholds.
6. Risk mitigation strategy and ALARP[2] demonstration.

The QRA is a screening level analysis that requires a comprehensive review of the potential failure modes to identify the ones that can cause loss of containment scenarios during the transportation operations (Ref. 5). Typically, the scenarios can be categorized as:

- External impacts (e.g., crashes, derailments, dropped objects, etc.)
- Spontaneous mechanical failure of the container device or its components
- Escalation scenarios (e.g., pressure relief device releasing due to external fire)

The identification of scenarios to be considered in a transportation QRA is done through the development of an event tree. Figure 1, an example event tree for a TIH transportation QRA, shows that scenarios from the categories above are included. Also, the event tree includes the different parameters of the scenarios to create a comprehensive mapping of the potential outcomes. Representative release sizes, weather conditions, and wind directions must be included in the QRA.

Once the scenarios are identified, their consequences can be calculated using dispersion modeling. Figure 2 shows example toxic dispersion contours for a scenario at a given point of a route using BakerRisk's proprietary software SafeSite$_{3G}$©. The concentrations calculated from the dispersion calculations are then used to calculate the lethality of the toxic cloud, typically using probit models. Since the scenarios can occur at any point of the transportation route, the consequence analysis must consider the different population numbers along the route as well as the relative location of the populations with respect to the potential scenarios. Population data is typically derived from census data or other population estimates to calculate the potential number of fatalities of the scenario. This thorough approach is computationally intensive and requires software tools able to handle thousands to millions of scenarios, depending on the characteristics of the route.

[2] As Low As Reasonably Practicable (ALARP), UK Health and Safety at Work etc. Act 1974.

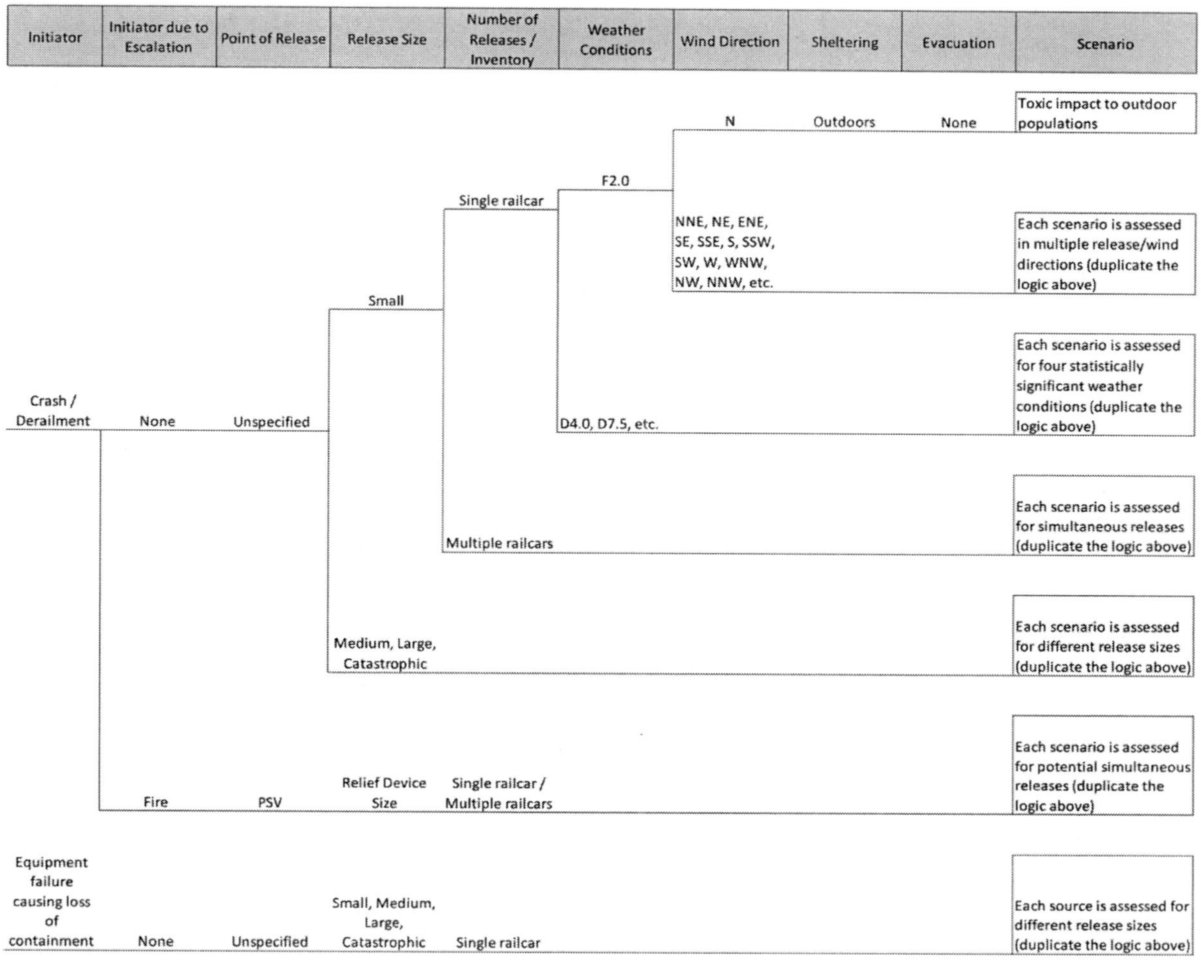

Figure 1. Example Event Tree for TIH Transportation QRA

Figure 2. Example Toxic Dispersion Contours

A frequency analysis is required to calculate the risk of each scenario once the consequences are calculated. In the frequency analysis, statistics on transportation incidents are used to estimate the likelihood of each scenario identified in the event tree and conditional probabilities are applied to the different parameters of the scenario (e.g., release size, weather condition, wind direction, etc.). However, it is important to understand the basis of the statistics reported as well as accounting for historical data specific to the operation analyzed.

The risk results can be presented in many ways, such as FN curves, societal risk, and location specific individual risk. However, the most typical way of reporting transportation risk is using FN curves based on the risk tolerance criteria lines for transportation of dangerous substances provided by the UK Health and Safety Executive Advisory Committee on Dangerous Substances (HSE ACDS) (Ref. 6). Risk tolerance criteria allow the owner/operator to classify the risk results based on published risk tolerance from which it can be determined if risk mitigation is required.

Figure 3 shows example FN curve results. It is observed that there is intolerable risk (above upper criteria line) for the baseline transportation operations; this requires the implementation of risk mitigation to decrease the risk to tolerable levels. Once the baseline risk results are obtained, the QRA model can be used to quantify the effectiveness of risk mitigation options. The risk mitigation strategy should be developed following the risk management hierarchy in Table 1. For example, alternative routes through locations with less population, enhanced design of the containers that reduce the likelihood of a release upon impact, controls above the industry norm, etc. can be evaluated using the QRA model. Risk reduction of each mitigation option is calculated and compared. In the example in Figure 3, the Risk Mitigation Option 2 reduces the risk more significantly than Option 1. Using the risk mitigation results with the estimated cost of implementing each of the risk mitigation options allows for a cost-benefit analysis to assist in the decision-making process.

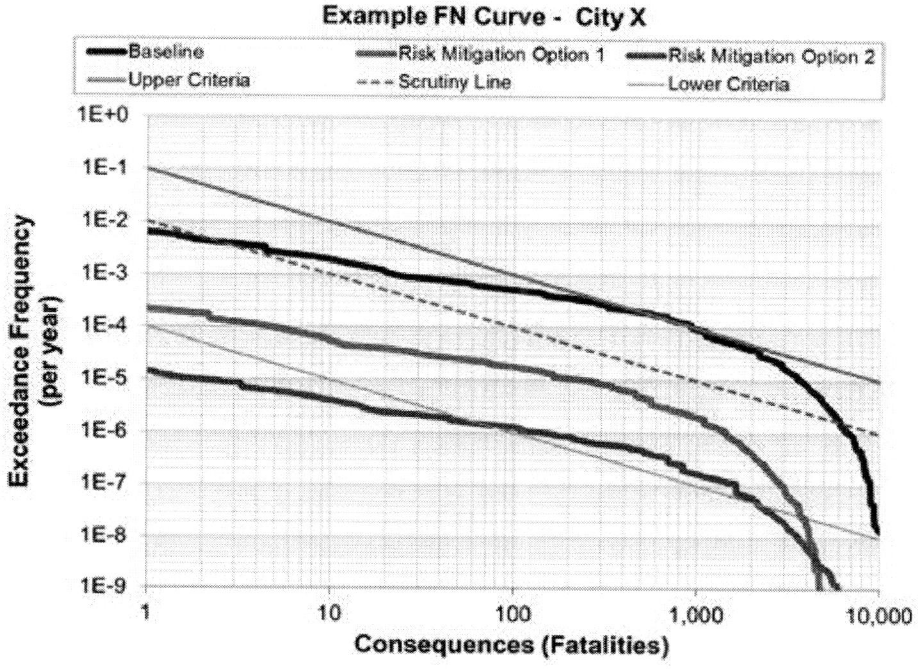

Figure 3. Example FN Curve Results[3]

4 Audit Protocol for Toxic Inhalation Hazard Chemicals

Auditing is a critical evaluation of a facility's management system to identify and correct weaknesses in the management system design and implementation. Auditing can be applied to process units handling and transporting TIH chemicals to identify opportunities for reducing risk and improving the effectiveness of the management system. To truly manage TIH chemicals it is necessary to look beyond local regulations, such as OSHA PSM[4] in the USA. Compliance with regulations is no guarantee that process safety incidents will be prevented. Facilities need to (i) identify the hazards, (ii) understand the risks associated with the hazards, and (iii) manage the risks by "*doing what is right*". This invariably means going beyond regulatory compliance and a risk-

[3] Risk tolerance lines based 0n HSE ACDS guidance (Ref. 6).

[4] Occupational Safety and Health Administration, Process Safety Management of Highly Hazardous Chemicals standard (29 CFR 1910.119).

based approach (i.e., 20 elements of CCPS RBPS) to assess all hazards and risks that could lead to an onsite and/or offsite TIH incident may be necessary.

Not all hazards and risks are created equal. Therefore, the emphasis of the auditing should be on what is truly safety critical, and dig deeper where the risk is high. While all RBPS elements should be addressed, in the case of TIH chemicals the main audit focus may be on:

- Operations personnel's understanding of hazards and risks,
- Operations and maintenance personnel competencies (knowledge, skills, abilities, and experience – not just training),
- Maintaining primary containment in pressure vessels, piping (including cross-country pipelines), tankage (including ISO tanks), road tankers, and rail cars,
- Safety culture and conduct of operations.

The author has developed a detailed and comprehensive audit protocol specifically for process facilities (process units and transportation) that handle TIH chemicals. This protocol addresses all 20 elements of RBPS, plus two additional elements for design and sharing learnings that are not explicitly covered by the CCPS Guidelines for Risk Based Process Safety (Ref. 2), see Table 3. For facilities that are subject to periodic compliance audits under OSHA PSM, the protocol is designed to remove unnecessary duplication.

Table 3. Example of Audit Protocol for Design of Facilities Handling TIH Chemicals

Question	Thought Prompts
When designing new projects and MOCs, does the facility adopt an Inherently Safer Design (ISD) approach?	• Intensification – minimize inventory of hazardous materials? • Substitution – eliminate hazards by substituting less hazardous materials? • Attenuation – minimize operating severity by avoiding extremely high pressures, high temperatures, or low temperatures? • Simplification – avoid unnecessary complexity / redundancy to minimize potential release sources and facilitate operator understanding? • Risk reduction measures – preference for Prevention before Control before Mitigation? • Reliability / availability measures – preference for Passive before Active before Administrative? • Other?
Do existing and new facilities have prevention systems to avoid hazards from being realized?	• Corrosion – use of corrosion-resistant alloys? • Joints – use of welded piping to minimize flanges, screwed fittings and couplings? • Rotating machinery – use of canned pumps? double seals? • Other?

The intent is that prior to the audit, the client will determine jointly with the auditors on which elements the main emphasis and focus will be placed. For example, the audit may address all elements, omit some elements, or alternatively devote less time to some elements deemed less relevant to the client's operations. These decisions on the scope of the audit should be based on risk and past experience (incidents, near-misses, prior audits, metrics, citations, etc.). As indicated above, it is likely priority/weighting could be placed on certain elements associated with understanding hazards and risks, managing risks, and softer issues related to culture and conduct of operations.

This protocol has been successfully used at several client facilities, and the client's employees have also been trained in the topic and proper use of the protocol. In fact, the author's colleagues were also trained to ensure a consistent application and interpretation of the protocol before auditing a client's facility. Like any review or audit, a written report is produced to cover the observations, findings, and recommendations where potential weaknesses have been identified. As this is not a regulatory compliance audit, some clients prefer to conduct the audit confidentially under legal privilege.

5 Conclusions

TIH chemicals are widely handled in process units and transported by road, rail, marine, and pipeline. Fatalities and serious respiratory impacts can occur through personnel exposure if these chemicals are not managed properly with a sound safety culture and sense of vulnerability.

While the workforce in process facilities is at risk, transportation poses additional risks as an accidental release can potentially expose a large public population. A comprehensive quantitative risk analysis allows identification of potential high-risk locations and operations, as well as to calculate the risk reduction of mitigation options, if needed. This powerful insight can be used to develop cost-effective risk mitigation and facilitate the decision-making process using a cost-benefit analysis.

Auditing both process facilities and transport operations with a specialized RBPS protocol can also be a powerful tool to identify management system weaknesses and improvements *before* a serious incident occurs. When the RBPS audit is combined with a comprehensive QRA, the information is available to proactively manage, prevent, and mitigate potential hazards associated TIH chemicals.

So, are you going to inhale or are you going to assess and reduce your risks? Then you'll never try it again!

6 References

[1] CSB (2019), "Toxic Chemical Release at the DuPont La Porte Chemical Facility La Porte, Texas", Report No. 2015-01-I-TX, U.S. Chemical Safety and Hazard Investigation Board, Washington D.C.

[2] CCPS (2007), "Guidelines for Risk Based Process Safety", Center for Chemical Process Safety, American Institute of Chemical Engineers, New York, NY.

[3] ARIA (2015), "Hydrofluoric acid, the worst of all acids?", ARIA News Flash, The Analysis, Research and Information on Accidents (ARIA) database, Bureau for Analysis of Industrial Risks and Pollutions, France.

[4] Dunning, A. E. & Oswalt, J.L. (2007), "Train Wreck and Chlorine Spill in Graniteville, South Carolina, Transportation Effects and Lessons in Small-Town Capacity for No-Notice Evacuation", Transportation Research Record, Journal of the Transportation Research Board.

[5] CCPS (1999), "Guidelines for Chemical Process Quantitative Risk Analysis", 2nd Edition, Center for Chemical Process Safety, American Institute of Chemical Engineers, New York, NY.

[6] HSE (1991), "Major Hazard Aspects of the Transport of Dangerous Substances", Health and Safety Executive, Advisory Committee on Dangerous Substances (ACDS), UK.

617617 Means to Achieve Backflow Prevention of Hazardous Chemicals from Process Vessels to Utility Pipelines and Its Importance during HAZOP Studies

Abstract:

Design of a chemical manufacturing facility is a challenging task because one has to ensure that every design feature has been carefully analyzed for Hazard and Operability (HAZOP). Along with the HAZOP studies of the main process components like reactors, crystallizers, distillation columns, etc., it is imperative to study the possible failures of the utility connections, which may get overlooked. A system is as safe as its weakest point of failure. A common failure is the back flow of hazardous chemicals into the utility connections. Utilities like nitrogen, steam, plant air, and deionized water are often directly connected to process vessels. The utility systems are generally common to the entire plant and so depending on the reaction conditions in an equipment, backflow may occur and contaminate other equipment. Different backflow prevention instrumentation is required based on the utility service. For instance, gaseous and liquid utilities require different backflow preventors. While choosing the material of construction of pipelines, specification brakes are placed at junctions where utilities are connected to the process vessels. This talk will outline various backflow preventor setups that have been successfully implemented in different chemical manufacturing facilities. Based on the severity of failure, the back flow prevention appropriate for different utility services will be discussed. Different kinds of back flow preventors, including differential pressure measuring instruments, flow transmitters, liquid trap with level switch, on/off valve with interlock activation etc., will be explained in detail and compared for their function and safety features.

SPRING21
+17TH GCPS
A Joint AIChE and CCPS Meeting

Further Lessons Learned on How to Efficiently Perform the Necessary PHA of Startup, Shutdown, and Online Maintenance

By: William Bridges
Process Improvement Institute, Inc. (PII)
1321 Waterside Lane, Knoxville, TN 37922 USA
wbridges@piii.com

PROCESS IMPROVEMENT INSTITUTE

Prepared for Presentation at
American Institute of Chemical Engineers
2021 Spring Meeting and 17th Global Congress on Process Safety
Virtual
April 18-22, 2021

Further Lessons Learned on How to Efficiently Perform the Necessary PHA of Startup, Shutdown, and Online Maintenance

By: William Bridges
Process Improvement Institute, Inc. (PII)

Keywords: PHA, Human Factors, Process, Hazard Evaluation, HAZOP, Safety, Factors, Human Error, Operations, Maintenance, Hazards, Process Safety, Review, HRA, PSM, Inspection, Programs, Reliability, Ammonia, Emergency, Recommendations, Checklist, FMEA, Instrumentation, Process Hazard Analysis, Risk Management, Initial, LOPA, Materials, Relief, P&ID, Risk Assessment, Investigatory, Layer of Protection Analysis, Risk-Based Process Safety, RBPS

Abstract

Hazard evaluations, also called process hazard analysis (PHAs) have been performed formally in gradually improving fashion for more than five decades. Methods such as HAZOP and What-If analysis have been developed and honed during this time. Some weaknesses identified 30 years ago still exist in the majority of PHAs performed around the world. Critically, most PHAs do not thoroughly analyze the errors that can occur during startup, shutdown, and other non-routine (non-normal) modes of operations; sadly the commonly used approaches for PHA of continuous mode of operation only find about 5 - 10% of the accident scenarios that may occur during startup, shutdown, and online maintenance. This is true even though about 80% of major accidents occur during non-routine operations. Instead of focusing on the most hazardous modes of operation, most PHAs focus on normal operations (e.g., HAZOP of equipment nodes). In a majority (perhaps more than 80%) of both older operations and new plants/projects, the non-routine modes of operations are not analyzed at all. This means that perhaps 70% of the accident scenarios during non-routine operations are being missed by those PHAs. **If the hazard evaluation does not find the scenarios that can likely occur during these non-routine operations, the organization will not know what safeguards are needed against these scenarios.**

One reason that many companies do not perform PHA of Procedures is because they believe the time required for such analysis will be excessive. This paper shows clearly the best practices for (1) screening and ranking which procedures/tasks are most critical for analysis, (2) deciding on the method to optimize the investment of time, and (3) streamlining the documentation of the results of the PHA of procedures. **Following the steps outlined in this paper reduces the time needed for PHA of procedures by about 40% without negatively effecting the number of accident scenarios found during PHA of procedures. The result should be more sites completing PHA of procedures.**

INTRODUCTION

Human error in research, design, construction, installation, operation, maintenance, manufacturing, inspection, management, etc., can be considered the cause of almost all industrial accidents. (Experts typically quote that about 85% of accidents are caused by human error, though some say that except for natural disasters this figure is 100 %.) However, simply attributing these incidents to "human error" without evaluating the root cause implies that the errors are inevitable, unforeseeable, and uncontrollable. Nothing could be further from the truth.

Human errors are sometimes mistakenly called procedural errors. This is not true any more than saying all equipment errors are due to design errors. People make mistakes for many reasons, but PII estimates that only about 10% of accidents due to human errors in the workplace occur because of *personal* influences, such as emotional state, health, or carelessness; most human error is due to weaknesses in the control of human factors. Over the past five decades of industry research and observation in the workplace on human error, we have come to know that human error probability depends on many factors.

These factors are described in more detail in *Human Factors Missing from PSM* [2]. Other papers provide much more detail on each human factor and the relative weighting of each.[2, 3, 4]

These human-error causes (human factors), which in turn result from other human errors, are all directly within management's control. When using human error data for controlling initiating events (IEs) and independent protection layers (IPLs), the site should ensure that the factors above are consistently controlled over the long-term and that they are controlled to the same degree during the mode of operation that the PHA, HAZOP, What-if, FMEA, or LOPA covers. For instance, if workers are fatigued following many extra hours of work in a two week period leading up to restart of a process, then the human error rates can increase by a factor of 10 times or 20 times during startup.[5]

Although this paper focuses on the requirements of a PHA (a redo or new hazard evaluation of an entire process), the approach is equally effective for other hazard evaluations such as preliminary and detailed design reviews (for new/revised processes) and large management of change hazard reviews. This paper builds on earlier papers from PII, most notably *Best Practices for Addressing Human Factors during PHAs/HAZOPs* (Bridges, Zahrani)[29].

INCREASING REGULATORY PRESSURE

Industry has taken some initiatives on resolving this problem. One initiative was to improve the focus on PHA of non-routine procedures as part of the update to "Guidelines for Hazard Evaluation"[1]. A new Chapter 9, Section 1 was added that necessitates hazard evaluations of all hazards of the process during all modes of operation. This textbook also explained why, when, and how to perform such analysis of step-by-step procedures.

Many companies have taken the initiatives to do the same, including about 20% of the largest chemical, petrochemical, and refining companies. But, the vast majority of companies who should be analyzing step-by-step deviations are not; and the major accidents continue to occur partly because of this. As a result, US regulators are beginning to increase pressure on regulated companies to perform PHA's of All modes of operation.

US OSHA Regulation and Enforcement:

The US OSHA PSM regulation requires PHA of all hazards during all modes of operation as well, and several key citations since 1990 have focused on PHA of non-normal modes.

- **Before there was a PSM regulation from US OSHA**, the agency published CPL 2-2.45 (Systems Safety Evaluation of Operations with Catastrophic Potential)[6]. In this guidance document, OSHA stated that a human error analysis should address:
    - *Consequences of failure to perform a task.*
    - *Consequences of incorrect performance of a task.*
    - *Procedures and controls to minimize errors.*[6]

  This approach is still the fundamental analysis method for PHA of non-normal modes of operation.

- **Phillips 66 "PHA" Citation —A citation with 566 instances was issued to Phillips 66 in Pasadena, TX, following their 1989 disaster** that killed 23 workers.[7] **The citation was related to a violation of the General Duty Clause (Section 5(a)(1) of OSH Act of 1970).** US OSHA cited Phillips against the General Duty Clause, since the PSM standard (29 CFR 1910.119) had not yet been issued. OSHA cited Phillips for not protecting its workers from hazards of fire/explosion by, among others, not performing a PHA that should have included an evaluation of the effect of design modifications on operator performance, and the identification of the source of observed human error and the identification of human factors that could result in incident event sequences. The citation stated, "**This review should result in a systematic listing of the (1) types of errors likely to be encountered during normal or emergency operation, (2) factors contributing to such errors, and (3) proposed system modifications to reduce the likelihood of such errors".**

The settlement agreement[8] between US OSHA and Phillips included the following requirements for process hazard analyses (PHAs) of the rebuilt and surviving units:

- "Phillips will analyze each process...and will include human factors analysis ... [and] will be ...led by an independent consultant."
  - William Bridges (of JBF Associates at the time, now with PII) led these PHAs. Before these PHAs began, OSHA, Phillips, and Mr. Bridges decided that the best approach for finding all human error scenarios was to perform a HAZOP of deviations of the steps for the procedures governing activities for startup, shutdown, and particularly online maintenance.
- "Phillips will provide OSHA an independent consultant's evaluation of the adequacy of its settling leg maintenance procedures performed while the polyethylene reactors are in operation..."
  - As part of the settlement to meet this requirement, it was decided by JBFA, Phillips and OSHA to perform a Human Reliability Analysis (HRA) of the Setting Leg online maintenance procedure, to ensure that the statistical risk of the accident recurring is less than the background risk of driving to work.

The PHA and HRA resulting from the Phillips settlement agreement is presented as a Case Study later in this paper for sake of clarity.

- **Paragraph (e) of the US OSHA regulation on PSM, 29 CFR 1910.119,[9] and similar requirements in US EPA's rule for risk management programs (RMP), 40 CFR 68.24,[10]** specifically require that PHAs consider and address hazards of the process, i.e., all hazards regardless of the mode of operation (routine or non-routine).
  - 29 CFR 1910.119(e)(1) states that the PHA, "shall identify, evaluate, and control **the hazards** involved in the process"
  - 29 CFR 1910.119(e)(3)(i) states that the process hazard analysis shall address "**The hazards of the process**".
  - 29 CFR 1910.119(e)(3)(vi) states that the process hazard analysis shall address human factors.
  - Appendix C to the OSHA PSM standard states that both routine and non-routine activities need to be addressed by the PHA of the covered process.

  There is no qualifier that limits the OSHA PHA requirement to only routine modes of operation. PSM requires that **all hazards** related to the process be addressed, regardless of the mode of operation or activity (routine or non-routine).

- **OSHA Inspection No. 103490306 (Nov 2, 1992).[11]** In the first major PSM inspection in 1992 using 29 CFR 1910.119, OSHA assessed a serious violation when the PHAs did not address "human factors such as board operator error, line breaking mistakes, and improper lockout and isolation of process equipment," all of which are errors originating from failure to either perform tasks or perform them correctly.

- **US OSHA published an internal document on Program Quality Verification of Process Hazard Analysis in 1993 (by Henry Woodcock, of OSHA).**[12] This document states that a PHA should include analysis of the "procedures for the *operation* and *support* functions" and goes on to define a "procedure analysis" as evaluating the risk of "skipping steps and performing steps wrong." The authors concur and PII has found the same true in PHAs that we have performed using various methods; a 2 Guideword HAZOP approach is normally optimal for PHA of procedures.

- **OSHA Inspection No. 123807828 (Nov 18, 1993)**[13] – Ashland Oil, Catlettsburg, KY. Several operators were preparing to ignite a 2-B-3 crude heater after a two week turnaround. The lead operator had two very inexperienced workers helping him light the heater. A large quantity of fuel gas entered the heater before the pilot light was ignited. The resulting explosion killed one employee, who received fatal injuries to the back of his head. The operators bypassed safety shutdown features; poor engineering allowed this to occur and should have been discovered in the PHA. In addition, they did not check the firebox to ensure that it was gas-free before lighting the heater.

  The Kentucky OSHA citation read: *The PHA did not address all hazards of the #2 Crude unit;.... The PHA did not address the hazards associated with the startup of the crude unit after a turnaround, ...emergency shutdown..., emergency operations and normal shutdown of the unit. The process hazard analysis that was completed by the PHA team for the #2 Crude unit only evaluated the hazards associated with normal mode of operation of the #2 Crude unit.*

  Settlement: All procedures were re-written and all PHAs were redone to include a PHA of deviations from procedural steps for all non-continuous modes of operation.

- **US OSHA PSM National Emphasis Programs** for Chemical Processes[14] and also for Refineries[15] underscore the need for companies to identify potential accident scenarios during non-routine modes, and to reduce the frequency and consequences of such errors as part of an overall process safety management (PSM) program.

OSHA recognizes that CCPS/AIChE has added as Chapter 9.1 in the 3rd edition of *Guidelines for Hazard Evaluation*[1] to further emphasize the need for a PHA to include hazard evaluations of all modes of operation and that this chapter has added best-practice detail on the approach for doing the hazard evaluation of startup, shutdown, and online maintenance modes of operation. Despite the specific OSHA standard that requires PHAs of covered processes must address all hazards, many PHAs still do not address hazards during all modes of operation. Further, many of the regulated community have stated "Well, OSHA did not tell us to perform a PHA of procedures for non-routine modes of operation." On the other-hand, OSHA did not state to do only a hazard evaluation of normal mode of operation and stop there.

To highlight the importance that PHAs address hazards during all modes of operation and activities (routine and non-routine), OSHA is considering issuing a Hazard Alert that would incorporate the concepts in Chapter 9.1 of *Guidelines for Hazard Evaluation Procedures, 2008, CCPS/AIChE.*[1] Also, as stated above, OSHA has an enforcement initiative, CHEM NEP, that utilizes a list of dynamic questions that OSHA compliance officers use to evaluate compliance at facilities covered by the program. **It is possible that future dynamic list questions could address PHAs of all modes of operation, and is further possible that this CHEM NEP update is drafted and waiting for release.**

Pressure from the US Chemical Safety and Hazard Investigation Board (US CSB)

The CSB has commented on the need for PHAs to address all hazards of the process during all modes of operation. Their clearest statement was in the report 2008-08-I-WV-R1[16] on the Bayer CropScience accident in Institute, WV, 2008. In that report, CSB asks Bayer to:

- Revise the corporate PHA policies and procedures to require:
  a. Validation of all PHA assumptions to ensure that risk analysis of each PHA scenario specifically examines the risk(s) of intentional bypassing or other nullifications of safeguards,
  b. **Addressing all phases of operation and special topics including those cited in chapter 9 of "Guidelines for Hazard Evaluation Procedures" (CCPS, 2008),**
  c. Training all PHA facilitators on the revised policies and procedures prior to assigning the facilitator to a PHA team, and
  d. Ensure all PHAs are updated to conform to the revised procedures.

US EPA's RMP Regulation

In the Risk Management Program rule (40 CFR 68.24)[10] EPA also recognizes the importance of procedural analysis, by defining the purpose of a PHA to **"examine, in a systematic, step-by-step way, the equipment, systems, and *procedures* (emphasis added) for handling regulated substances."**

> *A well-done PHA should identify all failure scenarios that could lead to significant exposure of workers, the public, or the environment.......For toxics under PSM, however, you may plan to address a loss of containment by venting toxic vapors to the outside air. In each circumstance, a PHA should define how the loss of containment could occur. However, for EPA, the PHA team should reassess venting as an appropriate mitigation measure. (From EPA RMP Guidance, Chapter 7, pgs 7-6 & 7-7; General Risk Management Program Guidance.*[17]*)*

Example of Local Regulation: Contra Costa County Hazard Materials Program

The counties in California are the implementation and enforcement agencies for the US EPA RMP regulation, which in California is termed, California Accidental Release Prevention (CalARP) regulations. One premier implementer is Contra Costa County (CCC). In addition to the standard requirements found in EPA's RMP regulation (which has requirements essentially identical to OSHA PSM), CCHMP has also added their own initiatives to improve how the 10 major facilities in CCC address human factors and PHAs of all modes of operation. The Industrial Safety Ordinance (ISO)[18] specifically requires that each site perform a PHA of procedures, just to be certain PHAs of all modes of operation are performed. One question in the county's auditing protocol is:

- Did the Stationary Source perform Procedural PHAs to evaluate potential active failures or unsafe acts in the procedure such as missed or out of sequence steps and including raising questions regarding the availability of personnel to perform a task as specified in the procedure? [Section B: Chapter 4.3 of the CCHMP Safety Program Guidance Document]

Conclusions on Regulatory Pressure

Clearly, the regulatory pressure is increasing for industry to perform a PHA that thoroughly addresses hazards during all modes of operation, including deviations from steps in startup, shutdown, and online maintenance procedures.

By the way, a similar focus is underway by the same government entities listed above to improve the coverage of all damage mechanism (corrosion, erosion, external impact, etc.) within a PHA. (such as CalOSHA's proposed rule for refineries). The 2008 update to the book *Guidelines for Hazard Evaluation Procedures*[1] was also to address weaknesses observed (across the industry) by US CSB (and others) in coverage of damage mechanisms within PHAs; US CSB requested these changes from CCPS.

Approach for Fully Addressing Human Factors in PHAs (Hazard Evaluations) - 4 Steps

To fully address human factors during PHA/HAZOP, a four-step approach is suggested.

- Step 1: Ensure education and experience of PHA/HAZOP Leaders in human factors and human error prevention. Also ensure the PHA/HAZOP Leaders are competent in PHA of procedures (which is very different than PHA of equipment nodes)
- Step 2: Ensure the brainstorming of scenarios includes consideration of human error and even multiple human error as causes; be specific as possible on the human error.
- Step 3: Have the PHA team perform a hazard review of procedure steps using a HAZOP or What-If analysis to uncover potential human errors associated with modes operations such as startup, shutdown, and online maintenance.

- Step 4: Supplement the PHA/HAZOP of the individual scenarios with a checklist analysis of general human factor issues, to ensure all major categories were addressed.

STEP 1 - Ensuring PHA/HAZOP Leaders are Competent in Human Factors and PHA of Procedures

The PHA Leader's understanding of human factors and human error prevention (HEP) will be the most significant indicator in whether the PHA will be successful in identifying the hazards caused by human error (the team will not be equipped to do this on their own, especially for scenarios that have not yet occurred). This understanding required of the PHA Leader is not a given trait that all leaders share (though it should at least be covered partially as part of PHA Leadership Training), it is a unique skill set that must be developed deliberately by the Leader, specifically by completing focused courses in human factors/human error prevention, participating in coaching by mentor/senior PHA Leaders, and studying the relevant statistics regarding human factors and human errors in industry. This training, along with several years in operation or safety roles to provide experience, will give the leader the ability to quickly recognize potential human factors (human error modifiers such as high fatigue or miscommunication) in the modes of operation being analyzed, and will provide the necessary tools to systematically brainstorm with the team in a way that maximizes hazard identification and meeting effectiveness. A large portion of these human error related hazards can only be identified with analysis of procedure steps (PHA of procedures, as part of the non-normal/non-continuous modes of operation), so the leader must be adept in techniques required for such analysis, including methods such as What If or 2 or 7 Guideword brainstorming of procedure step deviations, and the rules for analyzing and documenting procedure deviations and safeguards listed, discussed below in *STEP 3*.

STEP 2 - Ensuring PHA/HAZOP Leaders Look for and Find Human Error Causes during Routine and Non-routine modes of Operation

The training discussed above in human factors and human error prevention is mostly wasted if it is not applied to non-routine modes of operation (normal modes of operation generally are less dependent on human error). Many companies have not learned the basic truth that accidents that can occur during startup, shutdown, and online maintenance have little to do with accidents that occur during continuous more of operation. In fact, safeguards that adequately protect against accidents in normal mode, are of little or use for some scenarios that are unique to these non-routine modes of operation. Finding the unique safeguards necessary for the each mode of operation is critical. In addition to standard brainstorming of step deviations, the leaders and scribes most look for multiple simultaneous causes of an accident scenario (double jeopardy). This may seem like a conflict with traditional HAZOP rules, but in fact, there has never been an official rule against consideration of double jeopardy, though there has been recognition that none of the qualitative PHA methods are thorough in finding ALL double-jeopardy scenarios.

But during PHA of steps of procedures, it is fairly common for a double jeopardy to occur.

Therefore, the **company MUST require PHA of all modes of operation, and the analysis must be done the right way (as described in this paper).** This must be in a standard that is enforced by the company. At a minimum, the standard must:

1. Ensure the PHA Leaders are trained (Step 1) in the methods that best suit those modes of operation (like those in Steps 3 and 4), providing all procedures and related materials they will require (and ensuring that these materials are accurate and up to date), and that the PHA team includes the necessary expertise (experienced operators, senior operators/trainers as needed, etc.).

2. Ensure that the right methods are in fact used for leading and documenting the PHA of non-routine modes:
    a. This includes auditing the PHA meeting, as it is in progress, to ensure PHA is complying with Steps 3 and 4.
    b. Auditing the PHA report to ensure completion of the analysis per Step 3 and 4. Was there a PHA of steps? Were the right methods used? Is there evidence of unique safeguards being found for unique scenarios for each mode of operation? Is the evidence of discussion of double-jeopardy scenarios?

STEP 3: PHA of Non-Continuous Modes of Operation

During the period 1970 to 1989, 60% to 75% of major accidents in continuous processes occurred during non-routine modes of operation (principally startup and online maintenance modes).[19] This trend has continued unabated for most of the process industry to the present day. A compilation of 47 major process safety accidents from 1987 to 2010 was provided in an earlier paper on this topic (Bridges, et al) [20]; of these, 69% occurred during non-routine operations. In addition, a poll of over 50 clients indicates that 70% of their moderate and major accidents occurred during non-routine modes of operation. This data is particularly disturbing when factoring in the time at risk, since most continuous processes are typically shut down 5% or less per year. Therefore, for many continuous processes the workers and other stakeholders are 30 to 50 times more likely to have a major accident during the time frame of startup, shutdown, or on-line maintenance modes of operation.

One reason for processes being at higher risk during these operating modes is many of the safeguards (independent protection layers; IPLs) are bypassed or may not be fully capable in these modes. A hazard evaluation is necessary to help a company identify the layers of protection necessary to lower the risk to acceptable levels. To fulfill this need, a company operating a continuous process should **fully** evaluate the hazards during **all** modes of operation. Unfortunately, in the first four decades of hazard evaluation use (beginning after the Flixborough disaster in the UK in 1974 – an accident that occurred during startup in a temporary, poorly engineered configuration), many companies have done a poor job of identifying and evaluating accident scenarios during startup, shutdown, and online maintenance modes of operation, while usually doing a good job of evaluating hazards of normal modes (continuous or normal batch modes) of operation.

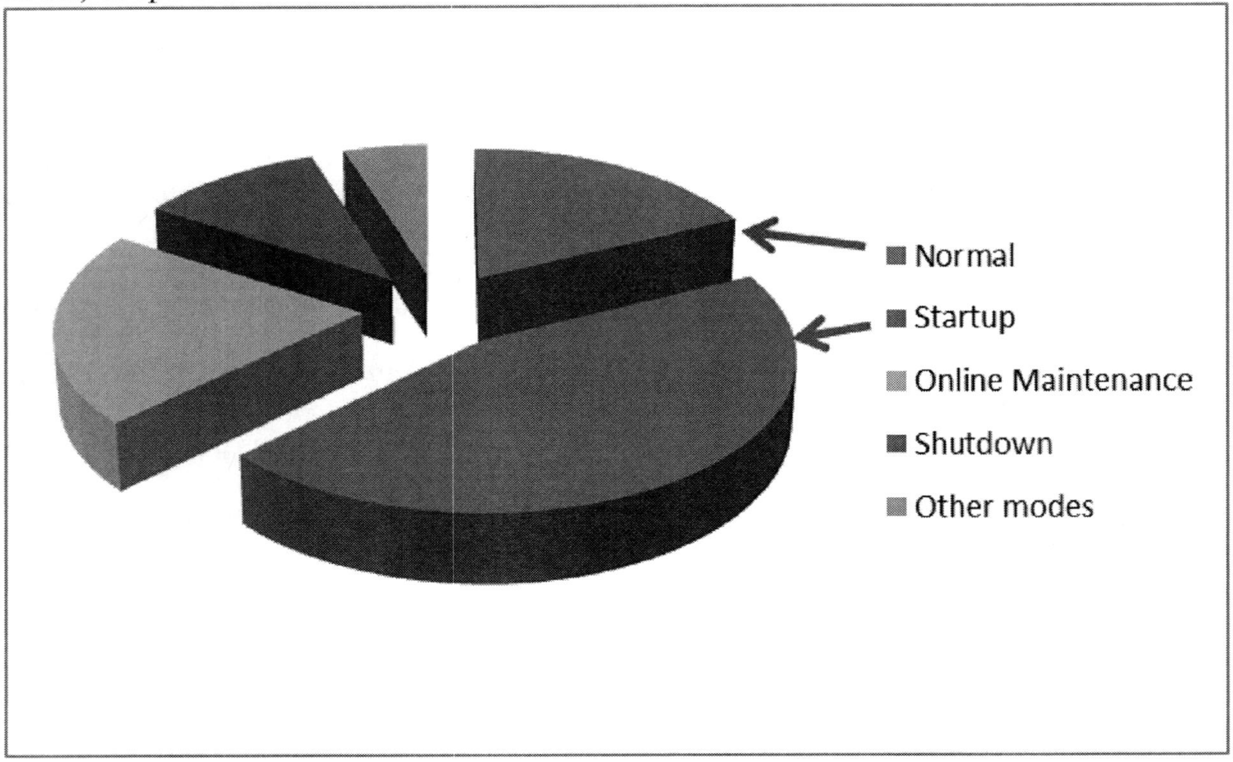

Figure 1: Distribution by Operating Mode of the 47 Largest Process Safety Accidents between 1987 and 2010[20]

How does someone responsible for coordinating or performing hazard evaluations (including PHAs) uncover potentially important accident scenarios during all modes of operation without consuming too many resources? To correctly answer this question, we must (1) understand the root causes of human error and (2) develop a strategy for systematically finding the scenarios that are caused by human error, during all modes of operation. The strategy must be thorough, yet provide for a practical allocation of resources. This paper provides a strategy that uses widely accepted hazard evaluation techniques (such as those referenced by OSHA and EPA for PHAs, which include what-if analysis and hazard and operability [HAZOP] analysis). This strategy has proven effective for hundreds of facilities over the past two decades since it was first published.[21] In addition to identifying accident scenarios during non-routine modes; this approach helps to more fully address human factors, which is a specific requirement of OSHA's PSM regulations and EPA's RMP rule.

Human factor deficiencies can make operations during non-routine modes extremely hazardous –

since operators generally have less operating experience for non-routine modes, and these types of operations rely heavily on operator decision-making and tasks. In addition, there are usually less layers of protection in effect during non-routine operations. Analyzing procedure steps can identify steps where the operator is most likely to make mistakes and suggest ways to reduce risk of an accident scenario, ranging from adding hardware to improving management systems.

The approach outlined in this work applies equally to any hazard evaluation where the steps for a non-routine mode of operation are well defined (i.e., written), including PHAs of existing units, hazard evaluations during preliminary and detailed design phases of projects (for new/revised processes), and large or small management-of-change hazard reviews.

Overview of Methodology for Hazard Evaluation of Non-Routine Modes of Operation

The hazard evaluation of non-routine modes of operation involves reviewing procedures using a HAZOP, simplified HAZOP, or What-if analysis to uncover potential accident scenarios associated with non-routine operations, for continuous or batch operations. As mentioned earlier, human error is more likely and more critical during non-routine operations. By analyzing procedural steps where human error is more likely, and where human error or component failure could lead to a consequence of interest, risk can be reduced. The objective for the hazard evaluation team is to evaluate the risk associated with skipping steps and performing steps incorrectly.

FMEA cannot be applied to procedure-based deviations, unless you create a "human" component, in which case you have simply merged HAZOP deviations for "steps" into FMEA. Pre-Hazard Analysis (P_rHA) and other hazard evaluation methods are not applicable for accomplishing a detailed hazard evaluation of non-routine modes of operations.

Checklist of human factors issues (see an earlier paper[21] and also Guidelines for Hazard Evaluation Procedures[1]) can be very useful after the detailed hazard evaluation of deviations of steps. Such analysis can indicate where generic weaknesses exist that can make errors during any mode of operation more likely, or that can make errors during maintenance more likely. Such human factors checklists are normally used at the end of the analysis, they can be done piecemeal during an analysis (on breaks from the meetings) by individuals on the team, and then the results of each individual review can be discussed as a team at the end.

As with scenarios uncovered during continuous modes of operation, the company may need to perform analysis (including semi-quantitative analysis such as LOPA or HRA) to more fully address any unresolved or complex issues raised in the hazard evaluation of non-routine modes of operation.

Case studies presented later in this paper illustrate the analysis approach and the usefulness of this strategy.

Purpose of Hazard Evaluation of Procedures-Based Modes of Operation

Although incorporating human factors considerations into hazard evaluation studies of continuous operation is straightforward by asking why the human might make a mistake that leads to a parametric deviation, this approach only addresses a small fraction of the potential human errors that can affect process safety. Many analysts have tried to find accident scenarios in non-routine modes of operations by adding generic guide words such as "deviations during startup" and "deviations during maintenance/sampling" to the hazard evaluation of equipment nodes/sections. Unfortunately, this only catches a fraction of the accident scenarios that can occur in non-routine modes since a hazard evaluation team is focused on "continuous" mode of operation during HAZOP or What-if of equipment sections/nodes.

From an informal survey of more than 100 companies, most do not currently perform process hazard evaluations of procedures, although many do perform some type of job safety analysis (JSA). The JSA is an excellent starting point for an evaluation of procedures because a JSA identifies the tasks that workers perform and the equipment required to protect workers from typical industrial hazards (slips, falls, cuts, burns, fumes, etc.). Unfortunately, a typical JSA will not usually identify process safety issues or related human factors concerns. For example, from a JSA perspective, it may be perfectly safe for an operator to open a steam valve before opening a feed valve; however, from a process safety perspective, the operator may need to open the feed valve before the steam valve to avoid the potential for overheating the reactor and initiating an exothermic decomposition; a PHA of procedural step deviations would likely find this deficiency, but a JSA will miss this process safety issue. The primary purpose of a JSA and other traditional methods for reviewing procedures has been to ensure that the procedures are accurate and complete (which is required of employers in 29 CFR 1910.119(f)(3)).[9]

By contrast, the purpose of a hazard evaluation is ***not*** to ensure the procedures are accurate and acceptable, but instead, to ***evaluate the accident scenarios if the procedures are not followed***. Even the best procedure may not be followed for any number of reasons, and these failures to follow the prescribed instructions can and do result in incidents. In fact, in the chemical industry and most other process industries the chance of an operator or other worker making a mistake in following a procedure is greater than 1/100, and in some cases much greater. When considering common human factor deficiencies that accompany non-routine operations, such as fatigue, lack of practice, the rush to restart and return to full production, etc., the probability of errors can climb to 1/10 chances per task (a task being about 1 to 10 detailed steps).[5]

The purpose of a hazard evaluation of non-routine modes of operation (governed by written procedures) is to make sure an organization has enough safeguards for the inevitable instance when a step is either performed wrong or skipped (inadvertently or due to shortcutting or other reasons)

Industry has found that a HAZOP or what-if analysis, structured to address procedures, can be used effectively for finding the great majority of accident scenarios that can occur during non-routine modes of operation.[1, 21, 22, 23] Experience shows that reviews of non-routine procedures have revealed many more hazards than merely trying to address these modes of operation during the P&ID driven hazard evaluations.

> ***Example:*** *For the BP Texas City, Texas refinery, pre-2005, if the isomerization unit had a hazard analysis of the startup mode (using What-If and 2-Guide Word analysis [explained later in this work]), the team would have likely identified that the high-high level switch in the column was a critical safety device. They also may have recommended moving the switch to a location higher in the column and then interlocking the high-high level switch to shutdown feed to the column. However, the site performed a parametric deviation HAZOP of the equipment nodes which focusing on continuous mode of operation, and so the team decided that the high-high level switch was not critical since devices in upstream and downstream process units (during continuous operation) would indicate possible level problems in the column – and besides, the operator would certainly notice the high-level condition on the sight glass during the rounds twice per shift. Unfortunately, these are not necessarily safeguards during startup of the column (1) since the routine practice was to overfill the bottoms (raise the level above the upper tap of the level controllers transmitter and above the nozzle for the high-high level switch and (2) since swings in upstream and downstream units are expected (and so likely such swings would not have led to intervention by the operators of the other units).*

To reinforce the need for and to explain the method for analysis of deviations of steps in a procedure, Section 9.1 was included in the 3rd Edition of *Guidelines for Hazard Evaluation Procedures,* 2008[1]; this was one of the major changes to the hazard evaluation procedures.

HAZOP Method for Analyzing Deviations of Procedural Steps

The Hazard and Operability (HAZOP) method has two major variations; one for the continuous mode of operations (where the team brainstorms what would happen if there were deviations of parameters) and procedure-based (where the team brainstorms what would happen when the steps of a procedure are not followed correctly). The procedure-based variation of HAZOP is the oldest form of HAZOP (from ICI in 1960s)[24]. It was an expansion of a Hazard Evaluation method based strictly on asking:

- **What happens if the step is skipped?**
- **What happens if the step is performed wrong?**

In turn, the "pre-HAZOP" method for brainstorming accident scenarios from not following procedures (including because the procedure is itself wrong) is based on the understanding that human errors occur by someone not doing a step (errors of omission) or by doing a step incorrectly (errors of commission). So, simply asking what would happen if the operator omitted a step or performed a step wrong is one way to structure a hazard evaluation of a step-by-step procedure. (We will discuss the usefulness of this simple approach to hazard evaluation of steps later.)

Seven (7) Guide Word Method

In an effort to be more thorough, the inventors of HAZOP (at ICI) broke these two types of errors into subparts and agreed on using the following 7 Guide Words:

Omission:	**Skip (or Step Missing)** **Part Of**
Commission:	**More** **Less** **Out of Sequence** **As Well As** **Other Than** **Reverse**

In the early 1990s, the guide word Skip was augmented by adding the option of discussing "are there any steps **missing** from the procedure."[21]

To apply HAZOP to procedural steps for startup, shutdown, online maintenance, and other modes of operation, the facilitator (or team) first divides the procedure into individual actions. This is already done if there is only one action per step. Then, the set of guide words or questions is systematically applied to each action of the procedure resulting in procedural deviations or what-if questions. The guide words (or procedural deviation phrases) shown in Table 1 were derived from HAZOP guide words commonly used for analysis of batch processes. The definition of each guide word is carefully chosen to allow universal and thorough application to both routine batch and non-routine continuous and batch procedures. The actual review team structure and meeting progression are nearly identical to that of a process equipment HAZOP or what-if analysis, except that active participation of one or more operators is even more important and usually requires two operators for a thorough review; a senior operator and a junior operator.

For each deviation from the intention of the process step (denoted by these guide words applied to the process step or action), the team needs to dig beyond the obvious cause, "operator error," to identify root causes associated with human error such as "inadequate emphasis on this step during training," "responsible for performing two tasks simultaneously," "inadequate labeling of valves," or "instrument display confusing or not readable." The guide word *missing* elicits causes such as "no written procedural step or formal training to obtain a hot work permit before this step," or "no written procedural step or formal training to open the discharge valve before starting the pump."

Table 1: Definitions of 7-8 Guide Words for HAZOP of Procedure-Based Operations

Guide Word	Meaning When Applied to a Step
Missing (optional guide word)	A step or precaution is missing from the written procedure prior to this step (similar to "Out of Sequence", except the missing step is not written)
Skip (No, Not, Don't)	The specified intent of this step is not performed
Part-of	A portion of the full intent is not performed. Usually only applies to a task that involves two or more nearly simultaneous actions ("Open valves A, B, and C".)
More	Too much of the specified intent is done (does not apply to simple on/off; open/close functions); or it is performed too fast
Less	Too little of the intent is done, or it is performed too slowly
Out of sequence	This step is performed too early in the sequence
As well as	Something happens, or the user does another action, in addition to the specified step being done correctly (could be a short cut)
Other than (or Reverse)	The wrong device is operated, selected, read, etc., or operated in a way other than intended. Or the wrong material is selected or added. "Other than" errors always imply a "Skip" as well.

Two (2) Guide Word Method for Analyzing Deviations of Procedural Steps

A more streamlined guide word approach has also proven very useful for (1) procedures related to less hazardous operations and tasks and/or (2) when the leader has extensive experience in the use of the guide words mentioned previously and can therefore compensate for the weaknesses of a more streamlined approach. The two guide words for this approach (as defined in Table 2 below) encompass the basic human error categories: errors of omission and commission. These guide words are used in an identical way to the guide words introduced earlier. Essentially "omit" includes the errors of omission related to the guide words "skip," "part of," and "missing" mentioned earlier. The guide word "incorrect" incorporates the errors of commission related to the guide words "more," "less," "out of sequence," "as well as," and "other than" mentioned earlier. Note that these two guide words (Table 2) fill the basic requirements for a human error analysis as outlined in OSHA's CPL 2-2.45.[6]

Table 2: 2 Guide Word (Guide Phrases) for Modified-HAZOP of Procedure-Based Operation

Guide Phrase	Meaning When Applied to a Step
Step not performed	The step is not done or part of the step is not done. Some possible reasons include the employee forgot to do the step, did not understand the importance of the step, or the procedures did not include this vital step
Step performed wrong	The employee's intent was to perform the step (not omit the step), however, the step is not performed as intended. Some possible reasons include the employee does too much or too little of stated task, the employee manipulates the wrong process component, or the employee reverses the order of the steps.

Table 3: Example of 2 Guide Word HAZOP of a Critical Step in a Procedure

Drawing or Procedure: SOP-03-002; ***Cooling Water Failure***	**Unit:** HF Alkylation	**Method:** 2 Guide Word Analysis	**Documentation Type:** Cause-by-Cause

Node: 23	**Description:** STEP 2: Block in olefin feed to each of the 2 reactors by blocking in feed at flow control valves				
Item	**Deviation**	**Causes**	**Consequences**	**Safeguards**	**Recommendation**
23.1	Step not performed	Operator failing to block in one of the reactors, such as due to miscommunication between control room operator and field operator; or control valve sticking open or leaking through	High pressure due to possible runaway reaction (because cooling is already lost), because of continued feeding of olefin (link to 11.7 - High Rxn Rate; HF Alky Reactor #1/#2) High pressure due to high level in the reactor, because of continued feeding olefin (link to 11.1 - High Level; HF Alky Reactor #1/#2)	High temperature alarm on reactor High pressure alarm on reactor Field operator may notice sound of fluid flow across valve Flow indication (in olefin charge line to reactor that is inadvertently NOT shutdown) Level indicator, high level alarm, and independent high-high level switch/alarm	
		Operator failing to make sure bypass valve is also closed, since this precaution is not listed in the written procedure; or the bypass valve leaks through	High pressure due to possible runaway reaction (because cooling is already lost), because of continued feeding of olefin (link to 11.7 - High Rxn Rate; HF Alky Reactor #1/#2) High pressure due to high level in the reactor, because of continued feeding olefin (link to 11.1 - High Level; HF Alky Reactor #1/#2)	High temperature alarm on reactor High pressure alarm on reactor Operator skill-training requires checking bypasses are closed, when blocking control valves Field operator may notice sound of fluid flow across valve Flow indication in olefin charge line (but likely not sensitive enough for small flows) Level indicator, high level alarm, and independent high-high level switch/alarm	
		Operator failing to close low control valve manually from the DCS because the phrase "block in" is used instead of the word "close"	Valve possibly opens full at restart, allowing too much flow to reactor at restart, resulting in poor quality at startup and/or possibly resulting in runaway reaction and high pressure	Control room skill training requires always manually commanding automatic valves closed before telling field operator to block in control valve	37. Implement best-practice rules for procedure writing, which includes using common terms.
23.2	Step performed wrong	Operator closing the olefin charge flow control valves before shutting down the charge pump, primarily because the steps are written out of the proper sequence	Deadheading of charge pump, leading to possible pump seal damage/failure and/or other leak, resulting in a fire hazard affecting a small area (link to 5.12 - Loss of Containment; Olefin Charge Line/Pump)	Step 3 of procedure that says to shutdown charge pump The step to shut down the charge pump (Step 3) is typically accomplished before Step 2 (in practice)	41. Move Step 3 ahead of Step 2.
		Field operator closing both upstream and downstream block valves	Possible trapping of liquid between block valve and control valve, leading to possible valve damage (due to thermal expansion)	Field operator skill training stresses that only one block valve should be closed	

What-if Method for Analyzing Deviations of Procedural Steps

The What-if method for analyzing procedure-based modes of operations is free brainstorming without the aid (or constraints) of guide words. This method is described in detail the *Guideline for Hazard Evaluation Procedures (CCPS)*[1]. The hazard evaluation team using this method would read the procedure and then answer the question: "What mistakes will lead to our consequences of interest?" The team would list these mistakes and then brainstorm the full consequences, causes, and existing safeguards – the same analysis approach described for the guide word approaches mentioned earlier in this section. What-if brainstorming **is not** applied to each step of the procedure, but rather covers the entire task (procedure) at one time.

Choosing the Right Method for Analysis of Non-Routine Modes of Operation

Obviously the What-if approach takes far less time than the 2-Guide Word method, and the 2 Guide Word method takes much less time than the 7-8 Guide Word method of HAZOP of procedures. Experience has shown that hazard evaluation facilitators, newly trained in the three techniques above, tend to overwork an analysis of non-routine procedures, so a tiered approach is best. In this tiered approach, the first step in choosing the right method of analysis in the hazard evaluation of procedures is to screen the procedures and select only those procedures with extreme hazards. These procedures should be subjected to a detailed HAZOP analysis (7-8 guide word set) presented above. The 2-Guide Word set is efficiently used for less complex tasks or where the consequences are lower. The What-if method is applicable to low hazard, low complexity, or very well understood tasks/hazards.

Experience of the leader or the team plays a major part in selecting the procedures to be analyzed, and then in deciding when to use each guide word set.

Figure 2 shows the typical usage of the three methods described above for a typical set of operations procedures within a complex chemical plant or refinery or other process/ operation. Most of the procedures are simple enough, or have low severity hazards to warrant using the What-if method. Currently, the 7-8 Guide Word approach is used infrequently, since most tasks do not require that level of scrutiny to find the accident scenarios during non-routine modes of operations.

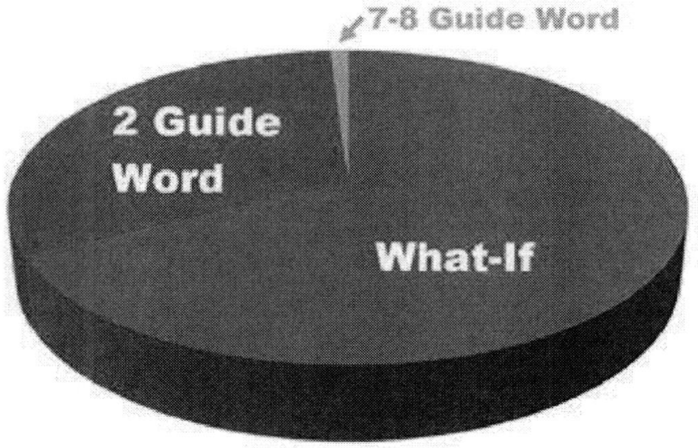

Figure 2. Relative Usage of Methods for Analysis of Procedure-Based Modes of Operation (essentially 2/3rd using What-if and 1/3rd using 2 Guide Word HAZOP)

The experience of the leader or the team plays a major part in selecting the method to use for each task/procedures to be analyzed. However the first decision will always be "Are these procedures ready to be evaluated to determine risk?" If the procedures are up-to-date, complete, clear, and used by operators, then the best approach for completing a complete hazard evaluation of All modes of operation, including routine modes of operation, is shown in Figures 3A and 3 B below:

Figure 3A: PHA of a ALL Modes of Operation for a CONTINUOUS Process

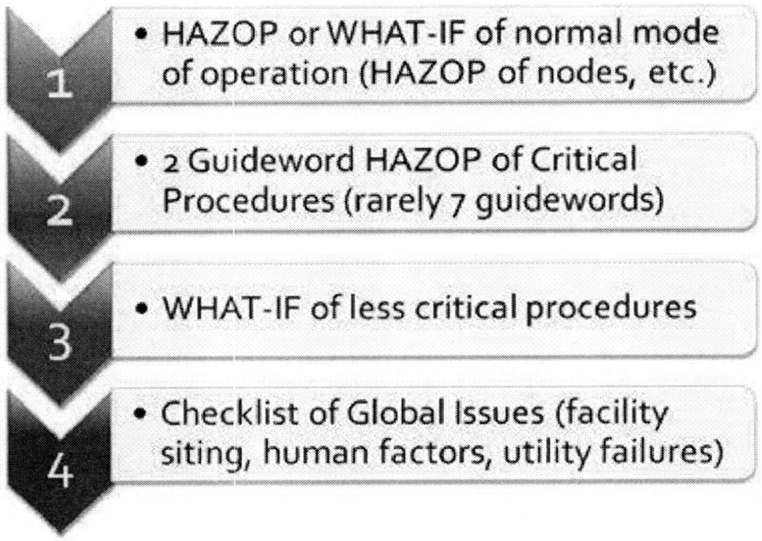

Figure 3B: PHA of a ALL Modes of Operation for a BATCH Process

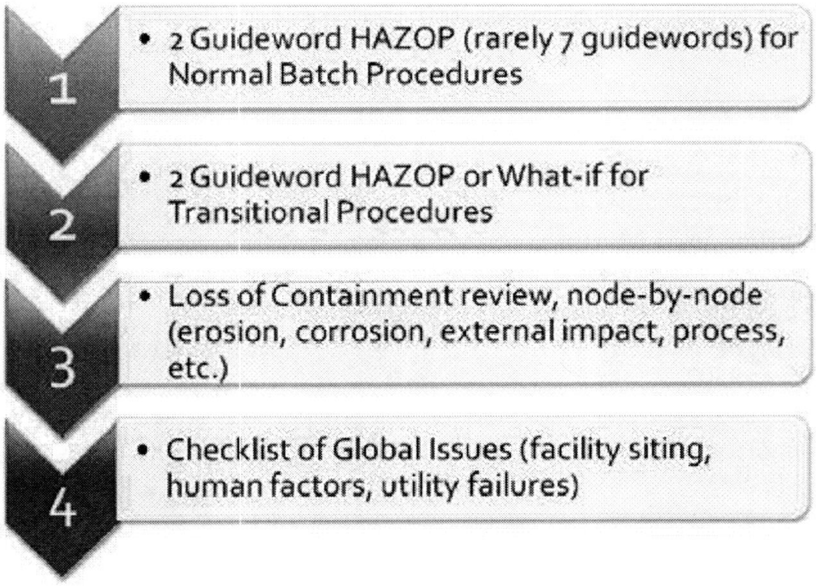

If procedures are not at least 90% accurate (with 95% accuracy being the target), then the best approach is to develop accurate and up-to-date procedures as quickly as possible and afterwards do a PHA of the newly issued procedures.

Any procedure (even a computer program) can be analyzed using these techniques. Reviews of routine procedures are important, but reviews of non-routine procedures are even more important. As mentioned earlier, the nature of non-routine procedures means that operators have much less experience performing them, and many organizations do not regularly update these procedures [though this should change as companies comply with 29 CFR 1910.119(f)][9]. Also, during non-routine operations, many of the standard equipment safeguards or interlocks are off or bypassed.

Using the approaches above, a company doing a complete hazard evaluation of an existing unit will invest about 65% of their time to evaluate normal (e.g., continuous mode) operation and 35% of their time for evaluating the risks of non-routine modes of operation.

Many companies do **not** perform a thorough analysis of the risk for startup, shutdown, and on-line maintenance modes of operation; the reason normally given is that the analysis of these modes of operation takes "too long." Yet, the hazard evaluation of the normal mode is taking too long and so the organization feels it has no time left for the analysis of procedures for startup and shutdown modes of operation. But, if these hazard evaluations for the normal mode of operation are **optimized** (such as using rules presented elsewhere[25]), the organization will have time for thoroughly analyzing the non-routine modes (typically discontinuous modes) of operation and the organization will still have a net savings overall! This point is critical since 60-75% of catastrophic accidents occur during non-routine modes of operation. Figure 4 illustrates (for a continuous process unit) the typical split of meeting time for analysis of routine mode of operation versus non-routine modes of operation.

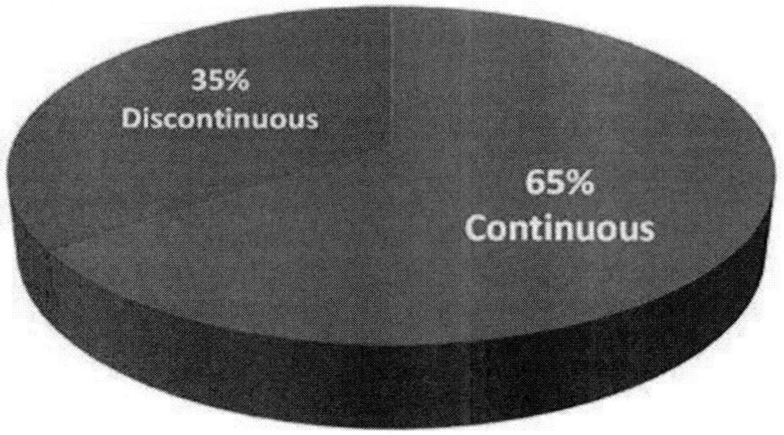

Figure 4. Relative Amount of Meeting Time Spent for Analysis of Routine and Non-routine Modes of Operation for a Continuous Process

General Guidelines for Analyzing Non-routine Modes of Operation or Batch (Step-by-step) Processes

Rank Procedures / Tasks. Together with an operator before the meeting or with the entire PHA team during the meeting, identify the sections of the procedures that warrant use of:

- 7-8 Guide Words (extremely large consequences can happen if deviations occur). Typically, in a chemical plant, petrochemical plant, gas plant, or refinery, no procedure will have hazards and complexity ***(High Complexity; Very High Hazards; such as explosives)***
- 2 Guide Words (the system is high to moderately complex, mistakes are costly, or severe consequences could occur) ***(High Complexity; High Hazards)***
- What-If (no guide words or guide phrases; for use on hazardous but simpler, less complex tasks) ***(Moderate to High Complexity; Moderate Hazards)***
- No detailed analysis (no further analysis for low hazard tasks) ***(Low Hazards)***

This is usually done by risk ranking the procedures High, Medium, and Low usimg the concepts of Hazard level of the task and Complexity level of the task. See Table 4 for an example scoring. Once the ranking is done, the PHA team then makes plans to do a PHA of the High risk procedures first, and then do the PHA of Medium ranked procedures. After some experience is gain, most PHA leaders find that 7-8 Guide Words is unnecessarily arduous and so the 2 Guide Word method is used to analyze the risk of deviations of steps of High risk tasks; and What-if is used for Medium risk tasks. The ones ranked Low are not analyzed further as the PHA team feels these tasks cannot lead to process safety scenarios.

Table 4: Example of Risk Ranking of the Procedures for Startup, Shutdown, and Online Maintenance

#	SOP #	Description	Rev #	Risk Ranking
1	BUT 900	Vaporizer (AE-1131/S) Preparation for Maintenance	00	H
2	BUT 901	Commissioning Process (Hydrocarbon) to Vaporizer (AE-1131/S)	00	M
3	BUT 902	Propane Refrigeration Compressor Motor, AKM-1160A/B Lube Oil System	10	L
4	BUT 903	Propane Refrigeration Compressor, AKM-1160A/B Start-up	11	M
5	BUT 904	Propane Refrigeration Compressor Normal Shutdown	10	L
6	BUT 906	Instrument and Shutdown Logic System for Chilled Water	10	L
7	BUT 908	Refrigeration Compressor Emergency Procedure Loss of Cooling Water / Instrument Air & Electric Power	11	L
8	BUT 911	C6 Transfer to De-ethanizer Bottom Steam Stripper (AC-0851 / Used Wash Oil Drum (AV-0841 / Flare K.O. Drum (Av-1161)	09	L
9	BUT 913	CAT-I Unloading from ISO- Container to AV-1165 to AV-1115	12	M
10	BUT 922	Unloading TEAL Catalyst Into Storage AV-1111 from ISO-Container	14	H

11	BUT 923	CAT-I / LC-2253 Catalyst Loading to AV-1115 from Barrels	10	
12	BUT 924	AMINE Loading from Drum	11	

On average we find that an optimum faction of procedures to review is about 20 to 30% of the total number of titles. Another rule of thoumb is to invest about half of the time spent on continuous mode PHA in analyzing SOPs to find the unique sceanrios that occur during startup, shutdown, and online maintenance. (So if the PHA of continuous mode required 10 days of meetings, then PHA of the High and Medium ranked SOPs for startup, shutdown, and online maintenance will require about 4 or 5 days of meetings.)

Details for PHA of Procedures (especially for PHA using HAZOP method)

- Define the assumptions about the system's initial status. “What is assumed to be the starting conditions when the user of the procedure begins with Step 1?”
- Define the complete design intention for each step. “Is the step actually 3 or 5 actions instead of one action? If so, what are the individual actions to accomplish this task?”
- Don’t analyze safeguard steps that start with ensure, check, verify, inspect, etc., or where the consequence of skip is “loss of one level of safeguard/protection against .....” There is no reason to analyze these steps since they will show up as safeguards of deviations of other steps. This approach is similar to not analyzing a PSV during a HAZOP of continuous mode (parametric deviation analysis); instead the PSV is shown as a safeguard against loss of containment.

- If the 2 Guide Word (or 7-8 Guide Word) method is to be used, then a couple additional analysis steps are needed:
  - Decompose each written step into a sequence of actions (verbs)
  - Apply guide words directly to the intentions of each action

The Following Preparation Steps May Also Be Needed:

- Walk through procedure in the plant with one or more operators to see the work situation and verify the accuracy of the written procedure. This is optional and should have also been performed as part of validation of the procedure after it was originally drafted.
- Determine if the procedure follows the best practices for “presentation” of the content; the best practices will limit the probability of human error.
- Discuss generic issues related to operating procedures, such as:
  - staffing (normal and temporary)
  - human-machine interface
  - worker training, certification, etc.
  - management of change
  - policy enforcement
- Review other related procedures such as lock out/tag out and hot work.
- **IF the procedures are NOT >90% accurate, then re-write the procedures first!**

STEP 4 - Using Checklist analysis to Help Ensure Thorough Coverage of All Human Factors

A general, template based Human Factors checklist questionnaire should be included in the scope of any large, unit-sized PHA to help ensure generic human factors are thoroughly considered. These checklists are available in the default templates of some PHA documentation software, such as LEADER ™ (from ABS Consulting). Even if not specifically called out in the PHA scope, it's worth going through template checklists during the last few hours of a meeting. If possible, the more efficient strategy is to send the checklists to team members a month or so before the meeting, so the team can go over the questionnaire, walk down the items on the checklists in the field (with an operator or other PHA team member, if possible), and then submit their responses to the PHA Leader prior to the meeting. Then, on the last day of the meetings, then Leader/Scribe reviews the consolidated list of issues found by the team members. This saves meeting time and allows for more in-depth consideration of the listed issues. (Note that generic checklists for Human Factors and Facility Siting are often completed together; and though the items listed on the siting checklist are not categorized directly as human factors, they include design considerations that may have been overlooked during the project/initial PHAs that can in turn lead to more human error.). Adding the Human Factors checklists to the analysis scope typically identifies more minor to moderate consequences (compared to those found in HAZOP of procedure steps or in HAZOP of normal mode of operation), and will generally yield far fewer recommendations (typically only 1-5% of the total recommendations come from use of the checklists). However, given the almost negligible additional costs to include the checklists in the analyses scope, the benefit from catching the last few hazards that might have slipped through the cracks otherwise will always justify such costs.

For an example of a ood checklist, see *Best Practices for Addressing Human Factors during PHAs/HAZOPs* (Bridges, Zahrani)[29], which can be found in the ***Free Resources*** section of www.piii.com.

CASE STUDIES

PII and others have performed tens of thousands of PHAs, including many thousands of PHAs of all modes of operation. Everyone who applies the lessons above find hundreds to thousands of scenarios that are unique to startup, shutdown, and online maintenance. For example:

> *Results from thousands of PHAs of non-routine modes of operation indicates that about 3% to 6% of the relief valves across the industry have been discovered to be too small and so had to be resized to account for accident scenarios that were only found during analysis of non-routine modes of operation.*

The following case studies illustrate the usefulness of the process outlined in this paper.

Case Study 1: Phillips Polyethylene Plant 6, Pasadena, TX

In 1991-1992, a PHA was performed for the first of the rebuilt polyethylene plants at the Phillips 66 plant in Pasadena, TX. The accident there two years prior claimed 24 lives, injured hundreds of others, destroyed all three polyethylene plants, and cost Phillips an estimated $1.4 billion. Following the investigation of the accident, one of the requirements of the settlement agreement between Phillips and the US government was to ensure the PHA of the rebuilt units addressed hazards during ***All modes of operation***.

The PHA team varied in size, but always included at least two operators. The team leader was a process engineer with 15 years of experience, who was also trained in human factors. The PHA first covered the continuous mode of operation for the approximately 200 nodes of equipment (from feed stock through pellet handling) using the "parametric deviation" form of HAZOP (and some What-If). Then, to complete the analysis of all modes of operation, the PHA team performed a step-by-step analysis of all steps of all startup and shutdown and online maintenance procedures (about 700 steps changed the state of the system and each of these steps were analyzed) using the 7 Guide Word HAZOP method (2 Guide Word analysis was not known to the team at this time). For deviations such as "operator skips a step," the causes identified by the team included "the operator doing this step miscommunicates with the operator who performed steps earlier in the day and went to the wrong reset panel/switch in the field". In this example, an "other-than" error led to the "skip" error; so two errors occurred at once: the wrong switch was flipped and the correct switch was not flipped. Other causes included: "label not distinct enough" or "thinking/believing the previous operator completed this step." The additional safeguards suggested by the PHA team sometimes lower the likelihood of the error by addressing a human factors weakness. But in many cases, the solution was a change to the hardware or instrumentation, including adding new interlocks (these would be called Safety Instrumented Functions today) and adding mechanical interlocks and installing larger relief valves. In a couple of cases, isolated sections of the process were redesigned to lower the inherent risk, such as adding error-proofing (Poke Yoke) features.

Figure 5: Typical Settling Leg Assembly for Phillips Polyethylene

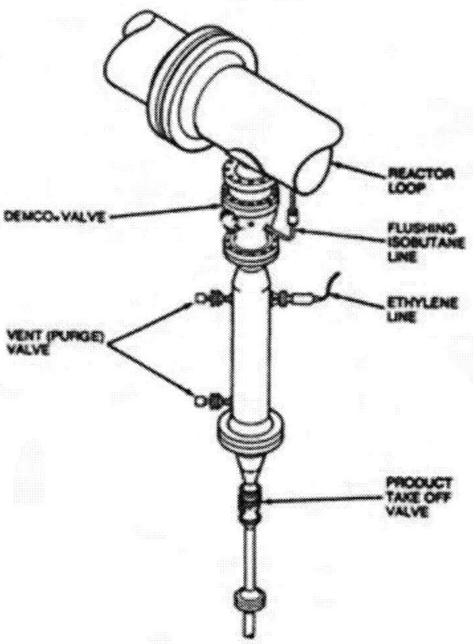

The 7 Guide Word HAZOP of non-routine modes of operation took 2.5 weeks of meetings, 40 hours a week. This was in addition to the 3.5 weeks of meetings to complete the parametric deviation analysis HAZOP of the continuous (normal) mode of operation (as mentioned before, 200 nodes of equipment); some all this the Normal PHA or Traditional PHA, but that is a misnomer. *Note that if the team had known of and been trained in 2 Guide Word HAZOP for procedure steps, they likely would have chosen that for many of the tasks and it is estimated that the meeting time for analysis of non-routine procedures would have been reduced to less than 2 weeks, with little or no loss of thoroughness.* The completed PHA report was submitted to US OSHA for review and

was approved almost immediately; OSHA particularly reviewed the analysis of all modes of operation and coverage of human factors.

After the PHA was completed, the settlement agreement also required a quantified human reliability analysis (HRA) of the online maintenance task of clearing a plugged settling leg – mistakes during this task led to the accident in 1989. The HRA was performed similar to those performed for nuclear reactor risk assessments, using a human reliability event tree to model the error probability for the task. The HRA results indicated that if the three IPLs in the new design are maintained, the probability of a similar event occurring when using this procedure was less than the risk of fatality when driving to work. This report was also approved by OSHA for settlement purposes.

The HRA shed light on new aspects of making errors and recovering from the errors during this task, but the HRA results did not result in changes to the process steps (at least not much), or the training program, or the human factors engineering, or the hardware/IPLs.

In this instance, the HRA validated the results of the 7 Guide Word HAZOP, but did not make new recommendations. The HAZOP of the procedures had already found the major accident scenarios and had already identified well enough the changes needed to reach tolerable risk.

Case Study 2: HUNTSMAN

Qualitative analyses of non-routine operating procedures is an extremely powerful tool for uncovering deficiencies that can lead to human errors and for uncovering accident scenarios during all modes of operation. Huntsman recognizes that regulators have required similar approaches for decades and that the regulators continue to note lack of analysis of the risk of non-routine operations and lack of risk review of changes to procedures.

The Huntsman Geismar site implementation of PHA of non-routine modes of operation in conjunction with the traditional PHA creates a complete process hazard analysis of all modes of operation. Using PHA of non-routine modes of operation, Site teams have already identified hazardous scenarios that exist only during startup that were not identified by the traditional PHA approach. Teams have also identified existing safeguards and potential new layers of protection to prevent the newly identified hazardous scenarios from occurring. The site has already installed inherently safer designs identified using the new methodology that significantly reduce the risk.

The Geismar site believes our implementation of PHA of non-routine modes of operation is a world class initiative that is necessary to identify all potential hazardous scenarios at our facility.

Case Study 3: UNITED (a SABIC affiliate)

In January 2019, PII completed a 'Re-do' PHA of the UNITED Ethylene Plant (Jubail, Saudi Arabia; a SABIC affiliate), which is to serve as that plant's new baseline PHA. This new hazard analysis included a PHA of Procedures, in compliance with SHEM 02.01, Rev 8, specifically section 5.12.2, which requires the PHA to consider all modes of operation, and section 5.12.2.9, that the PHA cover control system failures, including user interfaces and human factors.[26]

The meeting time was set at 19 days, with 14 days allocated for HAZOP of continuous/normal mode of operation and 5 days dedicated to PHA of Procedures (Step 3 of this paper) which was used to cover non-normal modes of operation: shutdown, startup, and online maintenance; and 3 hours for checklists reviews (such as Step 4 of this paper). For the continuous/normal mode of operation the plant was sectioned and analyzed in the typical HAZOP style, deviating each node's parameters as such as high and low deviations of level, flow, pressure, temperature, etc. as suitable for each node. The PHA of Procedures was done in the last 5 days, so the team was well aware of the major hazards and safeguards (at least for normal modes) relating to the equipment listed in each procedure. The procedure list was reviewed with the team on the first day to decide which procedures presented major process hazards (consequences of interest, in this case non-occupational hazards/serious injury or fatality consequences), so that more time could be focused on highest risks containing procedures. These procedures were typically more complex and usually longer in length. For these identified with significant process safety potential impact, the 2 Guideword Method was used and for those with less hazards, the What-If method was used. As usual, the goal was identifying specific scenarios of interest for those steps, capturing safeguards and safeguard steps in the documentation process. The few hours of the meetings included analysis using checklists to cover any hazards that weren't otherwise identified, and the team was given additional time outside of meeting to respond to the individual questions in both the Human Factors and Facility Siting Checklists, yielding an additional 3 recommendations deemed safety critical.

The PHA team identified many hazard during the meeting in both the normal mode HAZOP and the PHA of Procedures, listing 115 Safety Critical Recommendations (as defined by UNITED) and 7 Operability Recommendations (not safety critical, but have some impact to communication or effective operation). Of these Recommendations, 42 (or 36% of total) were identified during the PHA of Procedures; and while many of these were simple fixes needed to step order or wording,***14 were in response to critical (high risk) consequences identified in the procedure analysis, requiring new or upgraded independent protection layers*** to bring the risk to acceptable levels.

> *EXAMPLE:* During the procedure analysis for the acetylene reactors, it was discovered that there were no adequate safeguards against run-away reaction during start up, meaning the reactor shell could reasonably be expected to fail at some point due to human error during startup, which would likely cause a large explosion with the potential for multiple fatalities. The startup process required an extremely slow ramp-up in temperature (1°C per 5 minutes), and was controlled entirely by control room operators (by manually changing set points as they monitored for temperature spikes). In this case, the team recommended new

logic and safety instrumented functions to protect against the catastrophic reactor failure and explosion.

These hazards, many of which had NO safegaurds in place, would have been missed if the PHA scope failed to include procedure/non-normal modes of operation, which worth noting as statistically it is these types of errors/failures that lead to the majority of catastrophic consequences (non-normal modes of operation are when 80% of major accidents occur, as discussed previously). UNITED sees that the PHA of Procedures added extreme value to their current high value PHAs; and fully complies with corporate standards. As mentioned elsewhere, accurate procedures are needed to support a good PHA of procedures.

In summary, the PHA of procedures yeilded great benefit to UNITED by finding dozens of dangerous scenarios that were not found during PHA of normal mode of operation.

Another perhaps equally important aspect of human factors that was looked at carefully during the PHA was the fact that certain scenarios required more robust human intervention, esepcially when responding to some process alarms. These alarms are named "Critical Alarms" and directly trigger a Human Response IPL. These critical alarms required more attentive approach, where failure to respond adequately could results in severe consequences. Therefore, and in order to address the concerns of human errors that were specifically outlined in the work of Tew and Bridges (2010)[16], a common recommendation was made to UNITED to look at all alarms that are titled "critical" during the PHA study, and ensure that Training, Knowledge, and Skills are properly managed for these. For such Human Response IPLs, the pertinent operators must be provided a written trouble-shooting guide and each operator must be specifically trained for how to respond in time to each individual critical alarm. Moreover, response to these critical alarms must be tested and evaluated to ensure effectiveness of the Human Response IPL, and the time to respond (TTR) must be less than the safe span of time (within the Process Safety Time [PST]), based on simulated scenarios.

Another key human factor aspect covered in our PHA related to reducing specific human errors in the inspection and maintenance of Safety Instrumented Systems (SIS), which is a relatively new concept, largely introduced and proved by the work of Bridges and Thomas (2012)[28], and as it becomes increasingly important to maintain robust and effective high Safety Integrity Levels (high SILs). The PHA team closely looked at the concern of a possible increase in the likelihood of human error affecting the SIF's Probability of Failure on Demand (PFD). As a result, a recommendation was written for all of UNITED to consider maintenance staggering concept as a method to minimize the systematic probability of human error during proof testing and inspection. In essence, if a SIF has multiple channels (such as multiple sensors) that are voted 1oo2 or 1oo3, etc., then if the same person performs the calibration check of each channel, then the person should have 3 days separation between Channel 1 calibation and Channel 2, and similarly for Channel 3; otherwise the probability of repeating a failure on the 2nd and 3rd channel is about 90%. This is indeed an important consideration that needs to be always looked at when discussing human factors.

Case Study 4: SINOPEC-SABIC Tianjin Petrochemical Company (SSTPC)

SSTPC

The process plants at SS-TPC currently are:

- Ethylene (ET)
- MTBE & Butadiene (BD/MTBE)
- Phenol/Acetone (PHAC)
- High Density Polyethylene (HDPE)
- Linear Low Density Polyethylene (LLDPE)
- Polypropylene (PP)
- Pyrolysis Gasoline (DPG)
- Ethylene Oxide & Ethylene Glycol (EO/EG)
- Tank farm and Storage
- Utilities

PII led the PHA of the units, for all modes of operation. Besides a HAZOP or What-if of continuous modes of operation, the PHA team also used the Two Guideword or What-if approach to complete a PHA of startup, shutdown, and online maintenance modes of operation. The PHA of the non-routine modes of operation took about 20% of the total meeting time and was done at the end of the unit node-by-node analysis for continuous mode of operation.

Hundreds of scenarios were found during analysis of procedure-based modes of operation, resulting in many recommendations that had not been found during PHA of normal mode of operation. These resulted in implementation of new instrumentation and permissives and interlocks to reduce the accident scenario likelihoods.

Conclusions

Qualitative analysis of non-routine operating procedures is an extremely powerful tool for uncovering deficiencies that can lead to human errors and for uncovering accident scenarios during all modes of operation. This approach of step-by-step HAZOP and/or What-If analysis is not new to industry, and regulators have required similar approaches for decades. And regulators continue to note lack of analysis of the risk of non-routine operations and lack of risk review of changes to procedures.

> ***From the Wall Street Journal[87] referencing the presidential commission investigating the Deepwater Horizon accident of April 2010:*** *BP had rules in place governing procedural changes, but its workers didn't consistently follow them, according to BP's September [2010] internal report on the disaster and the report released earlier this month [January 2011] by the presidential commission on the accident. "Such decisions appear to have been made by the BP Macondo team in ad hoc fashion without any formal risk analysis or internal expert review," the commission's report said. "This appears to have been a key causal factor of the blowout."[27]*

> ***From CSB Report on August 2008 Bayer CropScience Explosion:***[16] *"The accident occurred during the startup of the methomyl unit, following a lengthy period of maintenance ... CSB investigators also found the company failed to perform a thorough Process Hazard Analysis, or PHA, as required by regulation...In particular, for operational tasks that depend heavily on task performance and operator decisions, the team should analyze the procedures step-by-step to identify potential incident scenarios and their consequences, and to determine if the protections in place are sufficient."*

More regulatory pressure is sure to follow, since major accidents continue to occur during non-routine modes of operation.

Regardless of what hazard evaluation technique is employed, it is imperative for PHA teams to ask, "Why would someone make this mistake?" whenever a human error is identified as a cause of a potential accident. "To err is human" may be a true statement, but the frequency and consequences of such errors can be effectively reduced with a well-designed strategy for analyzing risk of non-routine operating modes.

If the guidelines from PII outlined is Step 3 are followed, the PHA of startup, shutdown, and online maintenance will be state-of-the-art in efficiency and thoroughness, saving about 30 to 40% of alternate approaches to PHA of procedures.

References

1. "Guidelines for Hazard Evaluation Procedures, 3rd Edition, with Worked Examples," Center for Chemical Process Safety (CCPS), AIChE, New York, 2008.

2. Tew, R. and Bridges, W., "Human Factors Missing from PSM," *Loss Prevention Symposium (part of the Global Congress on Process Safety [GCPS]),* AIChE, March, 2010.

3. "The SPAR-H Human Reliability Analysis Method," NUREG/CR-6883, U.S. Nuclear Regulatory Commission, prepared by Gertman, D.; Blackman, H.; Marble, J.; Byers, J. and Smith, C., of the Office of Nuclear Regulatory Research, Washington, DC, August 2005.

4. "Human Event Repository and Analysis (HERA)," NUREG/CR-6903, U.S. Nuclear Regulatory Commission, prepared by B. Hallbert, A. Whaley, R. Boring, P. McCabe, and Y. Chang, 2007.

5. Bridges, William G. and Collazo-Ramos, Dr. Ginette. "Human Factors and Their Optimization." *8th Global Congress for Process Safety Proceeding (GCPS).* American Institute of Chemical Engineers (AIChE), 2012.

6. U.S. Department of Labor: Systems Safety Evaluation of Operations with Catastrophic Potential. Occupational Safety and Health Administration Instruction CPL 2-2.45, Directorate of Compliance Programs, September 6, 1988.

7. OSHA Inspection Number 106612443 - Phillips 66 Company, Houston Chemical Complex, Citations 1-1 through 1-566, Issued 4/19/1990.

8. *Stipulation and Settlement Agreement,* Phillips 66 Company ("Phillips") and Lynn Martin, Secretary of Labor, United States Department of Labor, 8/22/1991.

9. "Process Safety Management of Highly Hazardous Chemicals; Explosives and Blasting Agents," US OSHA Final Rule, 29 CFR 1910.119, February 24, 1992.

10. "Risk Management Programs for Chemical Accidental Release Prevention," US EPA Final Rule, 40 CFR 68, 1994-2000.

11. OSHA Inspection Number 103490306, Issued November 2, 1992.

12. Woodcock, Henry C., "Program Quality Verification of Process Hazard Analyses (for instructional purposes only)," US OSHA, 1993.

13. OSHA Inspection Number 123807828; Issued November 18, 1993.

14. U.S. Department of Labor: PSM Covered Chemical Facilities National Emphasis Program. Occupational Safety and Health Administration CPL 03-00-014, Directorate of Enforcement Programs, November 29, 2011.

15. U.S. Department of Labor: Petroleum Refinery Process Safety Management National Emphasis Program. Occupational Safety and Health Administration CPL 03-00-010, Directorate of Enforcement Programs, August 18, 2009.

16. "Investigation Report: Pesticide Chemical Runaway Reaction Pressure Vessel Explosion, at Bayer CropScience, LP, Institute, WV, on August 28, 2008", US Chemical Safety Board, Report No. 2008-08-I-WV, January 2011.

17. *EPA RMP Guidance, Chapter 7, pgs 7-6 & 7-7; General Risk Management Program Guidance.*

18. Contra Costa County Hazardous Materials Programs, *Contra Costa County Industrial Safety Ordinance (ISO),* by CCHMP. June 15, 2011.

19. Rasmussen, B. "Chemical Process Hazard Identification," *Reliability Engineering and System Safety*, Vol. 24, Elsevier Science Publishers Ltd., Great Britain, 1989.

20. Bridges, W. and Clark, T., "How to Efficiently Perform the Hazard Evaluation (PHA) Required for Non-Routine Modes of Operation (Startup, Shutdown, Online Maintenance)," *7th Global Congress on Process Safety [GCPS]),* AIChE, March, 2011.

21. Bridges, W.G., et. al., "Addressing Human Error During Process Hazard Analyses," *Chemical Engineering Progress*, May 1994.

22. Hammer, W., "Occupational Safety Management and Engineering, 3rd Ed.," Prentice Hall, 1985.

23. Kongso, H.E., "Application of a Guide to Analysis of Occupational Hazards in the Danish Iron and Chemical Industry," *International Conference on Hazard Identification and Risk Analysis, Human Factors and Human Reliability in Process Safety,* Center for Chemical Process Safety, AIChE, New York, 1992.

24. S. Mannan, ed., *Lees' Loss Prevention in the Process Industries, 4th Ed.,* Elsevier Butterworth-Heinemann, ISBN 978-0-12-397189-0, Oxford, UK, 2012.

25. Tew, R., et.al., "Optimizing Qualitative Hazard Evaluations (or How to Complete A Qualitative Hazard Evaluation Meeting in One-Third the Time Currently Required)," *5th Global Congress on Process Safety,* AIChE, April 2009.

26. *SHEM 02.01 Process Safety Risk Assessment, EHS-PRC-SM-002.01-07,* Rev 08, April, 2018, by SABIC

27. W. Bridges and H. Thomas, "Accounting for Human Error Probablity in SIL Verification Calculations" 8th *Global Congress on Process Safety*, AIChE, Houston, 2012.

28. *Wall Street Journal*, January 29, 2011.

29. W. Bridges and R. Zahrani, "Best Practices for Addressing Human Factors during PHAs/HAZOPs," 15th *Global Congress on Process Safety*, AIChE, New Orleans, 2019.

617764 Conducting Hazops That Are Good and Don't Just Look Good

Abstract:

It is difficult to imagine a time when HAZOP was not so prevalent in the process industry. But there was such a time. When HAZOPs first came about, they were performed much differently than they are today. Fast forward a few decades, and much has been gained in the implementation of HAZOP studies – *and much has been lost.* This paper gives a 40-year perspective on the history of HAZOP approaches and tools, including 32 years of the author's personal experience on how HAZOPs go right or go wrong. The pros and cons of the methods that are currently in use and misuse in different situations are discussed. The overall theme is to drive at the core of what constitutes a truly "good" HAZOP, rather than just performing a HAZOP that only *looks* good.

619090 Best Practices Driving Excellence in Plant Performance

Abstract:

Chemical and Process Plants are becoming increasingly more complex as smart equipment is becoming the new industry norm. Operators, Mechanics and Technical Personnel are constantly inundated with overwhelming streams of data. The growth of plant data has to some extent desensitized our staff to the intended warning flags through this data overload, unfortunately allowing critical information to pass through the system unnoticed. So, despite the sophistication of equipment and monitoring systems, plant failures are still too commonplace. As such, site performance continues to fall short of expectations as evidenced by unacceptable unplanned plant downtime, loss of containment related incidents, operator work-arounds due to equipment unavailability, and high loss profit opportunities. As expected, a retrospective analysis often in the form of event investigations most often reveal that existing plant data and "weak signals" were evident well in advance of the impending failures. Typically, readily available information and precursor events were either not detected, ignored or acted upon. Operating and technical personnel were unable to clearly identify, prioritize, communicate, and mitigate the growing plant risk to prevent the undesired "loss of control" and asset damage.

HOPE Consulting personnel with our broad experience from military, national research labs, commercial nuclear power generation, oil & gas, and mining industries has developed and implemented cross-industry best practices to address these performance shortfalls. The goal of our practices is to support uninterrupted runs between turnarounds, and greatly reduce the likelihood of significant equipment and human failures. These methods assist the operating staff to sort through the live plant data and plant physical anomalies to clearly identify, prioritize and mitigate impending equipment/plant failures. In addition, our process allows your organization to calculate the return on investments (ROI) resulting from our proven methods.

Through the integration of critical plant data, the operating teams can review & analyze performance, identify health threats & adverse trends, prioritize them appropriately and implement timely corrective actions. Team members are trained and become proficient in recognizing these "weak signals" and acting upon them before the equipment conditions become self-revealing e.g. equipment failure (enhancing equipment reliability). Personnel are trained in concept of "building the capacity to fail safe" increasing the likelihood that small failures do not cascade into major losses or incidents. Additionally operations, engineering, and maintenance personnel are mentored in the performance of Asset Integrity Inspections, a lost art as staff become increasingly dependent on process data and plant automation while becoming more disconnected from their plant equipment they maintain and operate.

Through this proven process, work teams have demonstrated their ability to proactively identify degrading conditions and underlying causes and enhanced prevention, rather than reacting to plant events. Additionally, business areas can now more effectively manage the supporting reliability programs and reduce the total number of exceptions and deviations which have contributed to unreliability in the past. Finally, systems and management processes are put in place to facilitate the necessary reinforcement, accountability, and process sustainability.

Keeping Process Safety Management Intact during Global Pandemic

Muhammad Salman Shafique
Fatimafert Limited
28-KM, Lahore- Sheikhupura Road, Lahore, Pakistan
Salman.shafique@fatima-group.com

Prepared for Presentation at
American Institute of Chemical Engineers
2020Spring Meeting and 17thGlobal Congress on Process Safety
Virtual
April 18 - 22, 2021

Keeping Process Safety Management Intact during Global Pandemic

Muhammad Salman Shafique
Fatimafert Limited
28-KM, Lahore- Sheikhupura Road, Lahore, Pakistan
Salman.shafique@fatima-group.com

Keywords: COVID-19, Process Safety Management, Crisis Management

Abstract

Process Safety Management (PSM) is the backbone of any process industry and is fundamental to its businesses, plant reliability, safe operations, employee well-being and prevention of catastrophic events.

In order to grow or sustain business and be responsible for its employees' health, good organizations never compromise on the foundation of PSM in their business and operational decisions. But what if a deadly disease alters the way how organizations and systems typically work? Can we quickly adapt to such unprecedented challenges? Are we able to fulfill all requirements of a process safety management system in this extraordinary scenario?

The paper will cover this topic in detail and it will shed light on the encountered challenges and tested best practices in order to maintain PSM during a global pandemic. Effects of a pandemic on key important factors related to PSM will be discussed like management of change, employee motivation, process hazard analysis, job supervision, process monitoring, competency levels, training & development etc.

Building on similar personal experience, the paper will also cover many interesting methods and suggestions for safety professionals and line managers to successfully adapt to the changes in this scenario while not compromising their process safety management systems.

1 Background

1.1 Introduction of Fatimafert Limited (FFT)

Fatimafert Limited is one of the three manufacturing plants of Fatima Group, which is the largest conglomerate of Pakistan producing Nitrogenous fertilizers along with businesses in Energy, Cement, Mining, Packaging and Commodities.

Fatimafert Limited (FFT) is located in the Lahore region of Punjab Province of Pakistan and produces Urea under the brand name of "Babarsher". Commissioned in 1971, it is one of the oldest plants of Pakistan producing Ammonia and Urea (prilled). It has 1000+ employees on permanent, company contract and third party contract basis.

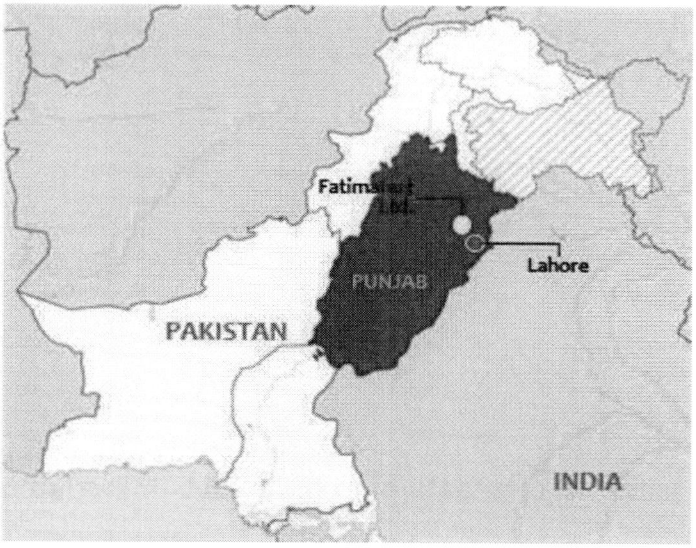

Figure 1. Location of Plant Site and Lahore in the map of Pakistan

1.2 Background of Process Safety Management at Fatimafert Limited

Fatimafert Ltd. has implemented DuPont's Process Safety Management (PSM) system which is a 22-element program consisting of 10 risk-based elements along with 12 cultural elements. All elements are driven by Visible Management Commitment and are connected with each other. Failure of any one element can result in a failure of the whole program. Fatimafert Ltd. has been declared on "compliance" level through a comprehensive assessment by DuPont.

1.3 COVID-19 Pandemic in the Region

COVID-19 pandemic rocked the world in early 2020 and was spread to Pakistan very quickly due to close proximity and continuous trade with China. First case of the disease was reported in Pakistan on February 26, 2020 and the country was swift in reacting to the pandemic. Lockdown was imposed and stringent SOPs were implemented by the government of Pakistan.

Fatimafert Ltd. plant site is situated near Lahore which is the 2nd largest city of Pakistan and is a sprawling metropolis with a population of around 12 million. As with all other densely populated cities of the world, Lahore remained the epicenter of the COVID outbreak in the region with highest COVID positivity ratio in all of the country.

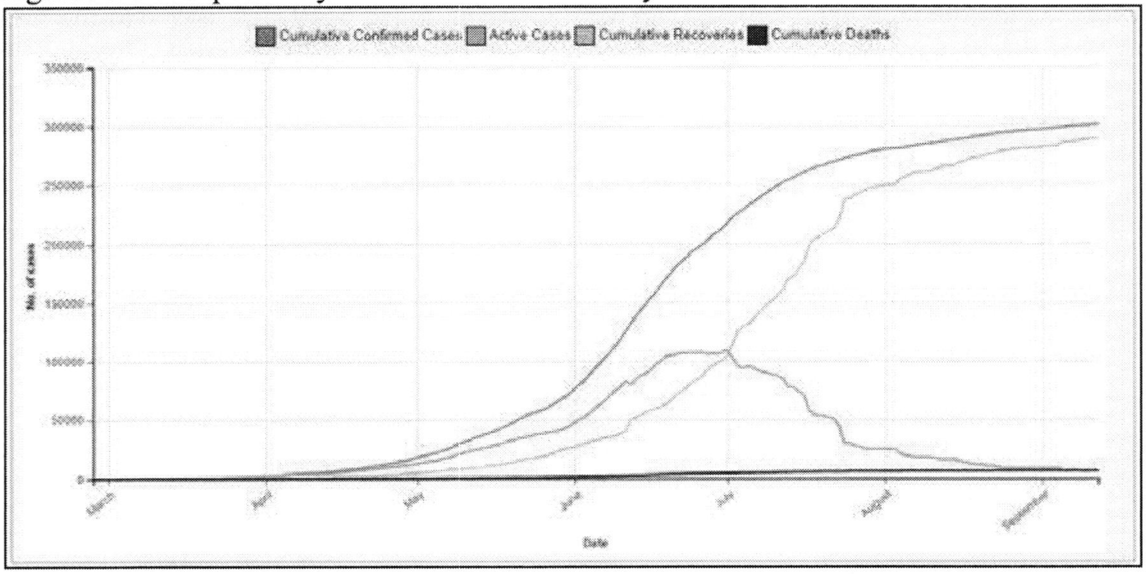

Figure 2. COVID-19 cases (cumulative) in first wave in Pakistan

2 Challenges due to Pandemic

Due to very close proximity with Lahore and almost 90% of workforce traveling daily between the plant site and different areas of Lahore, Fatimafert Ltd. faced a huge challenge to ensure the health and well-being of the employees and also to keep Process Safety uncompromised during operations of a highly hazardous fertilizer manufacturing plant. It was believed that any lapse in Process Safety Management system due to impacts of a very surprising and unusual situation can result in a catastrophic process safety incident and organization geared quickly to adapt according to the situation. Below are major challenges related to Process Safety which were encountered at Fatimafert Ltd. due to the pandemic;

1. Health & Safety of Employees

Keeping each and every member of the team safe from the outbreak remained one of the most pressing challenges as team consists of management, staff and contractual employees (skilled, semi-skilled and unskilled laborers).

2. Mental Well-being

A global pandemic and its rapid outbreak in the region seriously threatened to affect mental health of workers, which directly related with the cultural part of Process Safety Management and could have increased the process safety risks due to human errors.

3. Chaos due to Rumors and Inauthentic data

Due to the unique nature of the pandemic and many conspiracy theories associated with it, managing chaos and stresses due to inauthentic and over-inflated sensationalized statistics was a major challenge.

4. Operational Sustainability linked with National Food Security

Fertilizer production is directly linked with food security and keeping the national food supply chain intact in harsh circumstances remained a big challenge, which could only be achieved through uninterrupted and safe manufacturing operations.

5. Compliance to Process Safety Management

Usual 100% compliance of Process Safety Management program by all cadre of the workforce posed the biggest challenge as compliance levels in such unusual situations vary substantially from the routine safe circumstances. Along with the cultural elements, keeping the risk-based elements (e.g. Process Hazard Analysis, Process Safety Information, Management of Change etc.) intact was the biggest challenge due to abrupt changes to site dynamics including manning limitations, social distancing protocols, ban on physical meetings etc.

6. Minimum Competency Levels across functions

Management of the Personnel Change element of PSM requires every department to maintain a certain level of competency, which was challenged due to Work-from-Home or under-care employees and serious effects on conventional capability development program.

7. Field Supervision and Governance

Field presence and job supervision by operating personnel and maintenance supervisors were challenged due to work-from-home requirements and social distancing protocols.

8. Manpower Management during Lockdowns

Major and minor lockdown imposed by the government in severely affected areas of the region prohibited critical work force to travel and join plant duties putting safe plant operations in jeopardy.

9. Low Literacy related to COVID-19 amongst workforce

Overall literacy levels of the region are low and a new, unknown, widely-spreading disease further challenged comprehension capabilities of grass-root workers, making it tough for them to understand the risk completely and respond in a safe manner.

10. Gadgets and hardware for protection

Low comprehension of the disease, availability of personal protective equipment (PPE) and required protective gadgets remained a challenge especially during the first wave of the pandemic.

11. Risk Assessment & Hazard Analysis Studies

As risk assessment and process hazard analyses are a routine part of the hazardous process plants, organization of these multi-disciplinary team activities in restrictive conditions was very challenging.

12. Social distancing during operational and maintenance activities

Every facility, offices and buildings are designed to accommodate a specific number of people. But when the pandemic hugely limited occupancy levels, it put a question mark on safe execution of almost all operational and maintenance activities.

13. Budgeting and Funds for Safety

Extraordinary situations required extraordinary finances to overcome the challenges related to competency drainage, protective gadgets, testing, hospitalization without compromising routine safety related budgeting and expenses.

14. Securing the Employee Bubble instead of employee only

Due to high transmission rates of COVID-19, only putting efforts on keeping our employees safe was not sufficient unless the same level of efforts were put on awareness and security of families and social circles of the employees to curtail the spread.

3 Best Practices at Fatimafert Ltd. against Pandemic

Early realization of the extent of the threat the region faced and general risk awareness at Fatimafert Ltd. resulted in proactive planning and implementation of preventive actions including development of SOPs, protocols, employee education and handholding all segments of employees. COVID crisis management was formally triggered at Fatimafert Ltd. in last week of February 2020 after reporting of the first COVID positive case in Pakistan.

All efforts at Fatimafert were divided in *two major categories* as described below;

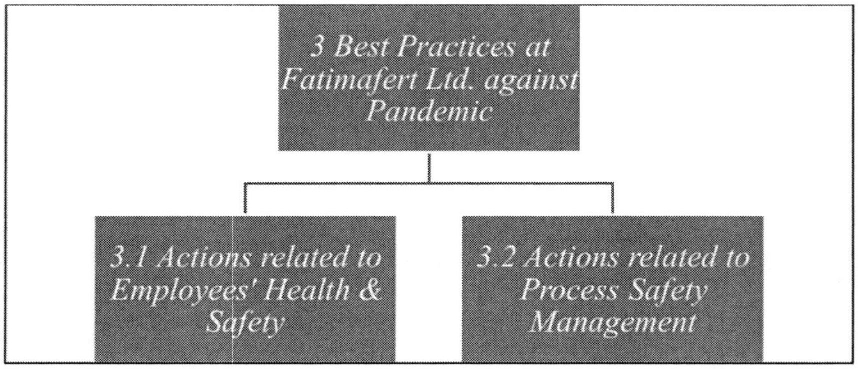

Figure 3. Major Categories of Best Practices at Fatimafert Ltd. against Pandemic

3.1 Actions related to Employees' Health & Safety:

Cultural elements of PSM are directly dependent on protection and well-being of workforce so dedicated efforts were initiated from the start to keep essential workers protected, safe and healthy from the disease. The COVID Crisis Management at Fatimafert Ltd. consisted of following four major elements;

3.1.1 Proactive Policies:

- **No Handshake Policy** – Handshakes and all types of physical contacts were immediately banned on company transport and within site premises.
- **Risk Profiling of Employees** – Medical risk profiling of all employees was carried out in order to clearly identify employees vulnerable to COVID infection including those with respiratory diseases and ischemic heart conditions.
- **Work-from-home for vulnerable employees** – Employees identified as "Vulnerable" through risk profiling were immediately barred from traveling to the site and were facilitated to work from their homes instead.
- **Shorter Working Hours for Immunity Boosting** – To facilitate employees (both traveling to site and on work from home) in having sufficient rest and to build immunity, duty hours were shortened significantly with same pay/wages.
- **Special Focus on Mental Well-being** – Emergence of COVID brought along with it huge anxiety and stress due to fear of the unknown. Therefore, the organization was quick in initiating special efforts to keep employees happy and relaxed. Special "Fun Breaks" were designed and incorporated in routine duty timings which included jokes, games, story sharing, physical exercises etc. In addition, a dedicated fitness program "Fit for Life" was launched during the first wave of the pandemic. Through this program, employees were coached and facilitated to develop a fitness mindset through steps tracking, weight loss, healthy diet, nutrient support etc. These efforts yielded surprising results and employees remained calm, relaxed and cheerful throughout the time of panic and despair.
- **Isolation of essential workforce in colony** – Fatimafert Ltd. maintains a small colony adjacent to the plant for accommodating critical workforce of all departments (around 50 families). Colony premises were locked down to ensure 100% security of the last line of defense; they were a critical workforce.
- **International/Domestic Travel and visitors handling protocol** – International and domestic traveling of employees were banned. A stringent health declaration and medical checkup protocol implemented for extremely unavoidable circumstances. All unnecessary incoming visitors and Vendor Services Manager (VSMs) were prohibited. Strict controls were implemented for extremely crucial VSMs only. These protocols

contained Declaration Forms (covering travel history, interactions history and family health status) and screening by medical professionals at the site's entrance.

- **Online meetings and trainings** – All team meetings and classroom training sessions were immediately prohibited and virtual meetings and training sessions were introduced. Required tools (software, licenses, user training) were immediately arranged for smooth execution.
- **Termination of gatherings & shutdown of Café** – All gatherings and team events were cancelled and café operation was terminated to eliminate the chance of cross contamination during meals preparation and consumption. All incoming employees were encouraged to bring along their meals from their homes.
- **Benchmarking of Policies with National & International Guidelines** – All COVID Crisis Management policies were benchmarked with best national and international guidelines including CCPS (Center for Chemical Process Safety), OSHA (Occupational Safety and Health Administration), government of Pakistan Guidelines & other national level manufacturing industries to learn from others.
- **Progressive Motivation** – In order to appreciate and encourage front line warriors like health care workers from medical center, plant operators, HSE (Health, Safety and Environment) Inspectors etc., a dedicated campaign by the name of "OUR HEROES" was launched. In addition, awards and recognition programs were also launched to acknowledge the efforts, reward the best contributors and keep the morale high.

3.1.2 Gadgets and Hardware:

- **Face masks and gloves for all employees** – N95 and surgical masks were arranged immediately for the whole workforce including contractual laborers and the use of masks in company transport and plant premises was made strictly mandatory.
- **Hand sanitizers & hand washing** facilities – Hand sanitizers were installed at entrances of all buildings and offices. Similarly, portable hand wash facilities along with soaps were placed at all entrances and places of usual commute of the workforce.
- **Extra vehicles for transportation** – Fatimafert Ltd. provides transportation to all outstation employees in the form of shuttle vans. In order to ensure social distancing (at least 3 feet) during commuting, vans occupancy limit was reduced to 50% of the total capacity and extra vans were arranged to ensure social distancing.
- **Full body disinfection at entrance** – Full-body disinfection enclosure were installed at all entrance gates of the plant site and colony.
- **Thermal screening of entrants** – Thermal screening devices were arranged and all incoming personnel were screened in two steps while entering plant premises. Anyone

having higher than normal body temperatures was denied access and escorted to the medical center for further checkup.

- **Installation of Door Openers** – Door openers were immediately provided to doors of all offices, washrooms etc. to keep them open and reduce chance of transmission through door handles.
- **Isolation Facility** – A secure isolation facility was established at a safe location to immediately isolate any employee showing symptoms during duty.
- **Anti-body test kits** – More than 100 Antibody test kits were procured for on-site verification of COVID infection in employees showing symptoms related to the disease.
- **Full body protection kits** – Full body protection kits were arranged for Health Care Workers of Plant Medical Center for safe medical screening of visitors and handling of employees with symptoms.

3.1.3 Employee Education:

- **Innovative Use of Social Media for Education of Employees** – In order to spread the awareness on authentic statistics and preventive measures to all level of employees down to grass root level, WhatsApp Broadcasting Groups were created and all company and contractual employees were added in those groups. Through it, only well-researched and reviewed information was shared with the workforce, which remained hugely effective and beneficial to avoid spread of rumors and panic situation.
- **Families & public education through Social Media Page** – A social media (Facebook) page was created specifically targeted to impart COVID education to employee families and the public. Education material shared through this page reached to residents of 70+ countries demonstrating a contribution of Fatimafert towards educating the community and the public in general.
- **Trends & Statistics** – Trends and statistics extracted of COVID information related to the region was compiled in easy-to-understand format and circulated across the site on a daily basis to inform employees on the ever-changing situation of the pandemic.
- **Communication circulars** – Communication circulars containing awareness data, preventive measures and company COVID related SOPs were developed in national language (Urdu) and circulated via emails as well as in form of pamphlets for distribution to employees and their families. These awareness brochures were also distributed in the nearby community.
- **COVID management certifications through online platforms** – COVID related awareness and management online training and certifications were arranged for employees from all departments who acted as COVID warriors in their respective domains.
- **Visualization of Preventions and Protocols** – All COVID preventive measures (e.g. hygiene etiquettes, hand washing method, relevant symptoms etc.) and company

protocols (where to report, how to report, SOPs, rules etc.) were designed in visually appealing patterns and clearly displayed at strategic locations in form of awareness boards, banners and posters.

- **Webinars on mental health** – Special webinars and coaching sessions were arranged for employees and their families by company medical officers on mental health, stress and anxiety management due to the COVID pandemic. It helped a huge deal in ensuring the mental well-being of employees and their loved ones.

3.1.4 Auditing:

- **Behaviors & awareness checks through Management Audits** – During MSAs (Management Safety Audit Program), special focus was put on verification and immediate correction of behaviors, compliance and awareness related to the COVID crisis management (e.g. masks compliance, social distancing compliance, physical gathering etc.). It helped greatly in aligning behaviors of the workforce with COVID protocols.
- **Reporting of Non-Compliances thru Observations Reporting Program** – Fatimafert Ltd. has a well-established and efficient medium for reporting of HSE observations by the name of "Mukhlis". All employees were encouraged and empowered to use this tool for reporting of any non-compliances of COVID protocols in company vehicles and plant premises. Best reporters were rewarded on their care for the colleagues.
- **Daily audits of face mask & social distancing compliance** – As face mask compliance was made 100% mandatory along with social distancing , a formal, round-the-clock auditing mechanism by a dedicated team to check compliance levels were successfully implemented.
- **Catering hygiene audits of workers canteen** – A workers' canteen was kept open under strict compliance watch to provide meals for shift working teams. Strict auditing regime was developed to audit COVID protocol compliance including social distance during meals collection, meals consumption, meal handling staff, etc.

3.2 Actions related to Process Safety Management:

Maintaining same compliance level of all Process Safety Management protocols and requirements during the pandemic was one of the biggest challenges faced at Fatimafert Ltd. Below is explanation of alternate methods taken against major elements of PSM;

- ### Risk Assessment & Process Hazard Analysis

Process Hazard Analysis is a team exercise and a discussion based brainstorming activity. Fatimafert Ltd. conducts the PHAs as per an approved PHA plan and PHA studies of three different process sections were due in 2020 as per the plan. It posed a great challenge to conduct the studies in COVID circumstances due to restrictions on physical meetings and most of the team members working from home. As a solution, remote PHA / HAZOP studies were introduced successfully. Area visits for facility siting survey and hazards identification were made shorter to reduce exposure time and were conducted by smaller group of PHA team members, staff interviews were conducted remotely and HAZOP study brainstorming sessions were carried out through virtual mediums (Skype for Business). Special trainings were arranged for PHA leader and team members for efficient utilization of time, resources and virtual mediums for this extremely important process safety activity.

- **Process Safety Information (PSI)**

Keeping Process Safety Information (PSI) include Hazards of Material, Process Design Basis and Equipment Design Basis. Keeping PSI documents of all units up-to-date and providing easy access to these packages for all employees in the office or working from home posed a major challenge and most of the documents (e.g. manuals, diagrams, flow-charts etc.) were kept in the Project Library. Therefore, after COVID arrived, extensive efforts were made to scan thousands of PSI documents and creating an online portal in which soft files of complete PSI packages were uploaded, sorted, catalogued and marked for quick and easy access by all employees from office and homes which they could use during PHAs, MOCs and various studies.

- **Management of Change (MOC)**

All non-critical and non-urgent modification jobs at the plant site were prohibited during the COVID pandemic. Only emergency changes were allowed after stringent reviews. Instead of paper-based forms and manual review process (office-to-office circulation of MOC documents posing risk of cross-contamination), a 100% digital, ORACLE-based platform was launched for virtual data entry, review and approval steps completion from laptops and PCs thus completely eliminating the need for physical review meetings and virus spread through paper-based files and documentation.

- **Mechanical Integrity & Quality Assurance (MIQA)**

Due to vintage design and aging of the plant, quality assurance, preventive maintenance of safety critical machines/equipment and their inspection becomes extremely important in order to ensure asset integrity and plant reliability. Due to this reason, the organization was not able to afford any delay or lapses in procurement/testing of safety critical devices and implementation of preventive maintenance/inspection plans. To assure quality of safety critical equipment being procured, all the correspondences were made online, virtual verifications during FATs (Factory Acceptance Test) were conducted instead of physical visits to the suppliers. A detailed SOP was developed for maintenance and inspection activities which chalked out stringent protocols related to tools sanitization, maximum technicians for each job, separate scheduling of each job etc.

• Incident Investigation & Reporting

In order to facilitate employees in immediate reporting of any Near Miss Event or any other type of safety incident and quality investigation/Root Cause Analysis by individuals and investigation teams, access of available online incidents portal Safety Management Information System (SMIS) was expanded, new licenses were purchased and reporting through paper-based forms were immediately prohibited to curtail virus spread. Through SMIS anyone could report an incident. Investigation teams were able to work together from their separate offices or homes through virtual mediums and conduct quality investigations. All interviews were conducted virtually and visits of incident locations were made shorter and by a maximum of two members of the investigation team following COVID protocols including masks and social distancing.

• Permit to Work (PTW) System

At Fatimafert Ltd. Permit to Work (PTW) system is a paper based system containing nine different types of permits including cold job, hot job, electrical job, instrument job, confined space entry, excavation, lifting over process lines, vehicle entry, and radiography permits. All kind of activities are covered through these permits and, due to complexity of the system and certain requirements, making the permits online was not possible. Therefore, alternate administrative controls were implemented instead. Instead of physical meeting of permit issuer and acceptor in field rooms, permit desks were set-up in open areas to ensure 100% ventilation, social distancing, and a queuing mechanism was introduced to reduce physical interactions. Instead of direct exchange of paper-based PTW forms between issuer and acceptor, PTW baskets were introduced to reduce chance of contamination and virus spread.

• Integrated Organizational Structure

Fatimafert Ltd. manages it Process Safety Management program through an integrated organizational structure of 13 committees including Executive PSM committee, contractors' management committee, behaviors and audits committee, occupational health committee etc. Each committee consists of 8-12 multi-disciplinary employees who meet at an identified frequency. Keeping in view high risk of contamination during these meetings, physical gatherings were immediately banned and virtual mediums (Skype for Business) were provided to all committee members for smooth meetings along with an online tool for recording, communication and stewardship of minutes of meeting.

• Audits & Observations

Process Safety Auditing Program consists of behavioral audits by site management called Management Safety Audits (MSAs) (60 audits per month), internal PSM audits (annual), second party audits by Fatima Group Corporate HSE team (annual) and external PSM audits by DuPont (usually after three years). As per plan all audits, except external PSM audit, were due during the pandemic and were conducted safely and successfully. SOP of MSAs was revised to incorporate guidelines for making safe safety contacts, shorter area visits and online audit report submission portal. On the other hand, preparatory, kick-off, discussion and closing meetings of all PSM audits were conducted virtually instead of physical sessions. All required documents and evidences were exchanged through shared folders instead of files and paper-based documents to eliminate risk of cross contamination.

- **Emergency Response & Planning**

Fatimafert Ltd. has defined roles and maintains a crisis handling team through two Emergency Control Centers (ECCs) for effective handling and control of any process safety emergency. Due to limits on site occupancy and work-from-home protocol implementation, the organization faces challenges related to availability and presence of key ECC members which could have affected gravely in case of emergency. To counter that challenge, back-up membership of all control centers was identified and socialized. Emergency role statements were revised accordingly and targeted training on these updated role statements were provided to new members. Moreover, weekly and the quarterly emergency drills/mock exercises mechanism was modified to meet COVID protocols including sanitization of emergency handling gadgets, fire and safety articles, identification of safe distances, group activities with a smaller workforce and critique meetings in open air instead of critique rooms.

- **Training & Development**

Due to the ban on physical / classroom training sessions, development of employees and their capacity building was a huge challenge which could have direct consequences on process safety of the plant. The company invested in purchasing well-recognized online training programs including Harvard ManageMentor, Linkedin Linda etc. Along with that, a dedicated "Capability Development Program" was launched and successfully executed through which employees of all cadres and groups achieved trainings and certifications on topics as diverse as COVID management, process safety, HSE, digital tools, remote working and functional / technical topics etc. It resulted in completion of 850 online courses (certificates & diplomas) resulting in 2230 training man-hours in addition to routine training man-hours derived from annual training calendars.

4 Conclusion

Despite of the very susceptible situation of Fatimafert Ltd. plant site due to proximity with the epicenter of the COVID outbreak in the region and very high exposure of 90% of employees to the affected areas, so far there have been a total of nine COVID positive cases in over 350,000 man-days. There has been no process safety incident at the site during the COVID pandemic, all process safety studies, audits and protocols are being followed as per normal conditions.

All of this had been possible just due to a proactive approach, flexible policies related to employee well-being but very strict protocols related to COVID crisis management, innovative methods of employee education (especially the use of social media), alternate methods to ensure compliance of Process Safety elements, digital tools, stringent auditing and, most importantly, care for employees and the community which reigned supreme during this time of need and mutual support.

SPRING21
+17TH GCPS
A Joint AIChE and CCPS Meeting

Pandemic Risk Management; Protecting people while ensuring business continuity

James Sneddon
Principal Consultant, Risktec Solutions (Canada) Ltd.
Suite 1000, Bankers Hall West, 888 3rd Street SW, Calgary, AB
James.Sneddon@risktec.tuv.com

Prepared for Presentation at
American Institute of Chemical Engineers
2021 Spring Meeting and 17th Global Congress on Process Safety
Virtual
April 18 - 22, 2021

Pandemic Risk Management; Protecting people while ensuring business continuity

James Sneddon
Principal Consultant, Risktec Solutions (Canada) Ltd.
Suite 1000, Bankers Hall West, 888 3rd Street SW, Calgary, AB
James.Sneddon@risktec.tuv.com

Keywords: Pandemic, COVID-19, Risk Management, Bowtie, LOD

Abstract

The COVID-19 pandemic swept across the globe in the latter half of 2019, throughout 2020 and into 2021. In response, many companies implemented work from home policies, while others stopped operations entirely in an effort to limit the spread throughout their workforce and supporting communities.

This containment strategy was not universally viable; long-term shutdowns impacted the economic viability of companies, and some industries were designated as an 'essential service' and thus continued operations. These employers faced the proposition of balancing the needs of the business and the community with a continued responsibility to provide a safe workplace for employees.

This paper, and accompanying presentation, will demonstrate how the application of bowtie analysis, a commonly adopted methodology in high hazard industries, can help the risks associated with continued operation in a pandemic to be better understood and managed, thus ensuring the safety of both your personnel and business.

1 Pandemic Risk Management; An Introduction

Since its appearance in December 2019, the SARS-CoV-2 virus (COVID-19) has continued to spread, touching almost all corners of the world. The progression from outbreak to pandemic was soon accompanied by the swift imposition of lockdowns in many countries.

In support of the various restrictions imposed upon their communities, many organizations implemented work from home policies, while others stopped operations entirely in an effort to limit the spread. There were, however, industries where this approach was not viable, either because they provided an essential service or because of economic necessity. Now, more than a year on from coronavirus initially upending daily life and work, this situation remains.

Employers requiring their employees to remain in, or return to, an office environment, work site or operating facility face the ongoing proposition of balancing the needs of the business and the community with a continued responsibility to provide a safe working environment for their employees.

Common risk management methodologies, such as bowtie analysis combined with the Lines of Defense (LOD) model, can be adapted to help employers better understand the risks involved with a resumption or continuation of in-person operations, ensuring the measures they have in place to manage this risk are appropriate and effective.

2 Bowtie Analysis

2.1 Introduction to Bowtie Analysis

Risk assessment lies at the heart of any form of risk management, and one of the most powerful of these techniques is the bowtie method. Its strength is that it goes beyond the usual risk assessment 'snapshot' and highlights the links between controls, assurance & verification activities, and the underlying management system; a valuable trait when assessing the constantly evolving nature of a pandemic.

Bowties originated as a method for assessing operational risk, with the earliest mention of such an approach appearing within an adaptation from the ICI plc HAZAN Course Notes 1979, presented by The University of Queensland, Australia [1]. The Royal Dutch Shell Group was the first major company to integrate the bowtie method into its business practices [2, 3, 4] and is credited with developing the technique which is widely used today.

The bowtie method provides a readily understood visualization of the relationships between the causes of business upsets, the escalation of such events, the controls preventing the event from occurring, and the mitigation measures in place to limit the business impact (see Figure 1).

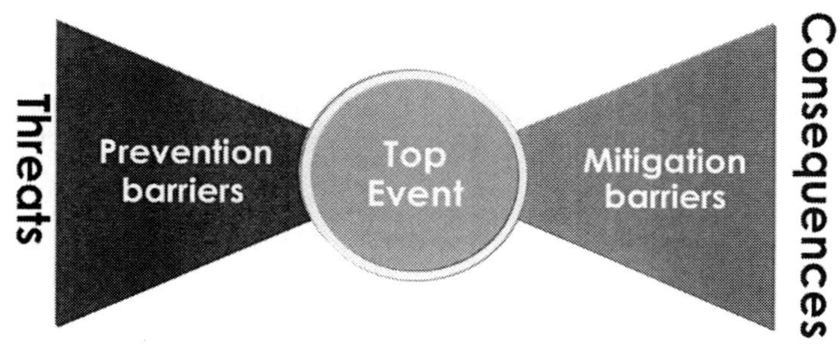

Figure 1: Bowtie Framework

A description of the different components of the bowtie, and the method for building such a diagram are well-documented [2, 3, 5], hence this paper focusses predominantly on its specific application within a pandemic risk management setting.

2.2 Application of Bowtie Methodology to Pandemic Risk Management

The application of bowtie methodology can be shifted from its more traditional use in high-hazard industries to assessing the risk to an organizations workforce during the COVID-19 pandemic in a relatively straightforward manner. Through graphical representation, bowtie analysis can map threats that may impact worker safety or the ability to maintain safe operation, identify and assess the safeguarding in place to prevent or mitigate different scenarios, and readily highlight any deficiencies or non-conformances.

A representative example of how the bowtie methodology can be applied in the assessment of 'worker safety' is provided in Figure 2, below.

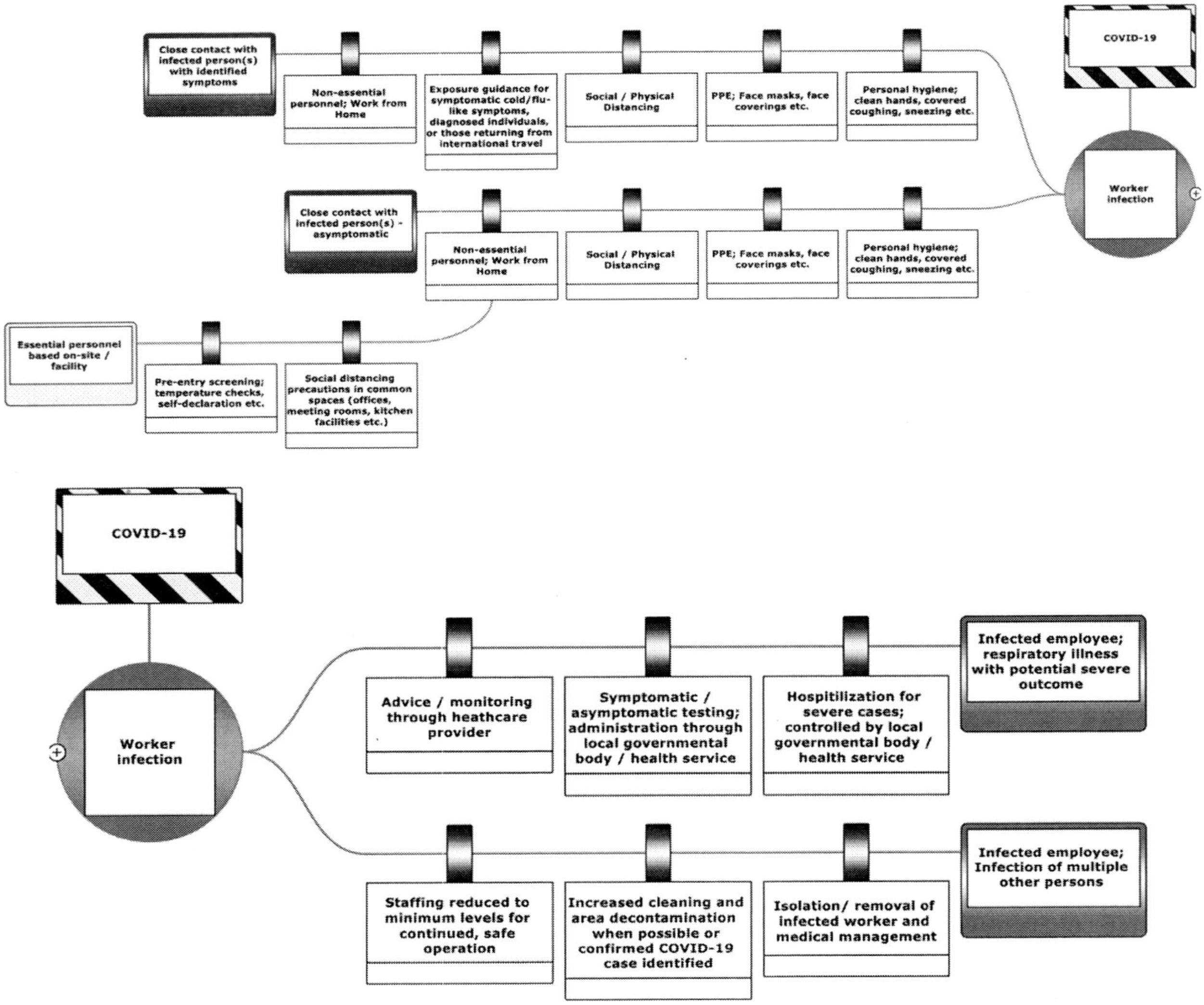

Figure 2: COVID-19 Worker Infection Example

3 Risk Management Framework

Bowtie analysis is not a panacea for managing risk in the current environment. It must be integrated within a wider risk management framework, enabling barrier implementation to be controlled and the risk profile for the organization to be consistently monitored.

The adoption of such a framework is of special significance to large organizations who face the following challenges:

1. How do we coordinate our response in a consistent manner across multiple facilities and offices?
2. How do we provide a large and geographically diverse workforce with readily available information on our response and the resources available?
3. How do we map and track compliance of safeguarding to ensure conformance and identify any deficiencies?

The following features of bowtie analysis, and its application within a robust risk management framework help address these challenges.

3.1 Pandemic Response Communication

The bowtie is an excellent communication tool for the coordination of response. Dissemination of key information to the workforce can be actively managed via targeted outputs from the bowtie diagrams.

Information for each barrier on the bowtie diagram can be easily and swiftly communicated to the workforce in a one-page summary (Figure 3), ensuring a coherent and well-informed response, covering questions such as:

- What is the barrier?
- What does it do?
- How does it perform?
- How is it tested?
- Where do I find documents with further information?
- Who should I contact for further details?

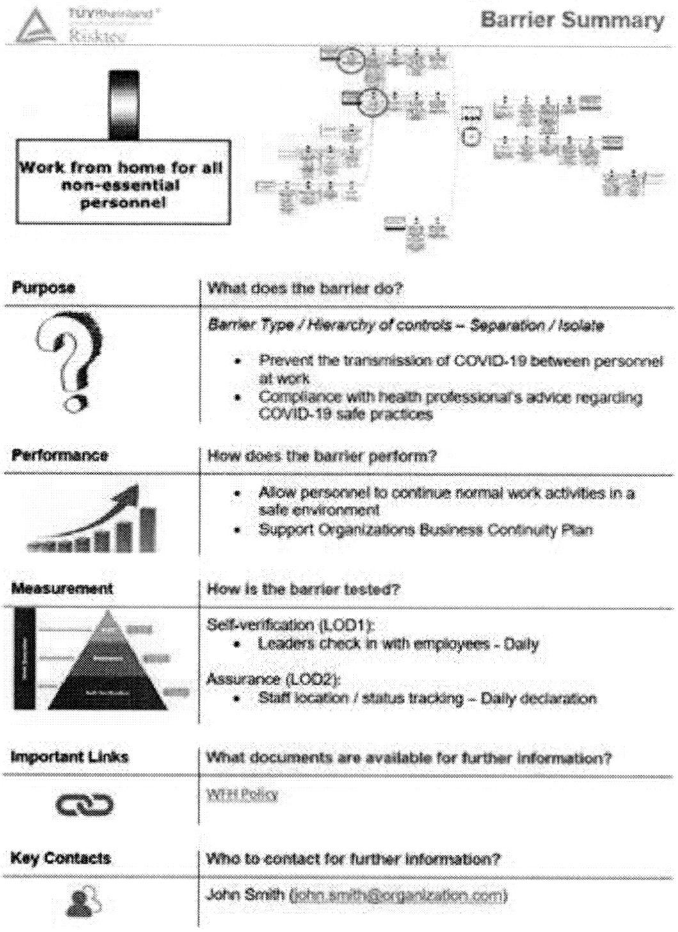

Figure 3: COVID-19 Safeguarding Summary

3.2 Barrier / Safeguarding Assurance and Verification

In order to track how well prevention and mitigation barriers are performing, assurance and verification criteria must be defined as benchmarks to measure success. Such criteria are commonplace within high hazard industries in the form of Performance Standards. However, a more appropriate approach for managing the COVID-19 response is the three lines of defense (LOD) model.

The LOD model is typically deployed in an internal auditing function [6]. LOD modelling is a method used to gauge performance, enhance clarity regarding risks and controls, and help improve the effectiveness of risk management systems. It is based around three broad concepts:

- LOD1: Self-verification that activities have been completed as prescribed. Barriers are effectively controlling risks and are delivering planned performance;
- LOD2: Independent functional assurance of conformance to requirements and quality of operating activities; and,
- LOD3: Internal, external and regulatory audits to confirm compliance.

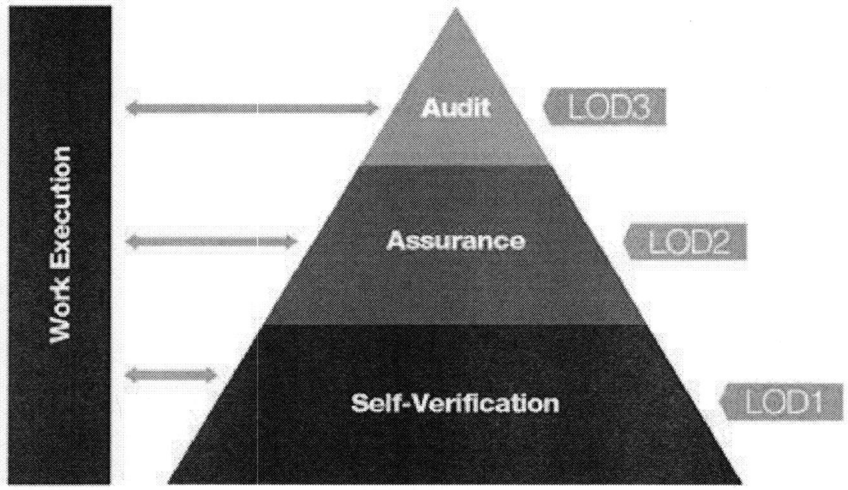

Figure 4: Lines of Defense Model

The LOD model can be applied to the prevention and mitigation barriers identified within the bowtie in the following manner:

1. Identify assurance and verification activities for each barrier;
2. Disseminate these criteria to operating locations, offices, entities in questionnaire format for completion by site representatives (see Figure 5);
3. Collate responses within the auditing function of the bowtie software used (e.g. BowTieXP); and
4. Display and monitor location performance using a Key Performance Indicator (KPI) dashboard (see Figure 6).

TÜVRheinland®
Risktec

Daily COVID-19 Checklist; LOD1 Self-Verification **Date:**

Barrier Ref.	Barrier	Self-Verification Action	Guidance	Responsible Person / Role	Check Status
W-P1	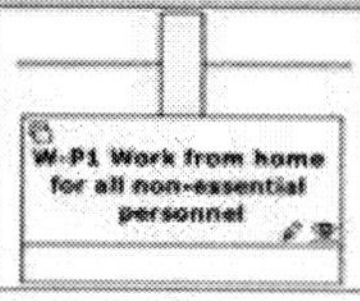	Check-in with 90% of employees & contractors	Check-in can be by video, phone, or e-mail. The 90% value is a guide to indicate that most staff should be regularly engaged, but 100% compliance is not realistic every day.	Department / Team Leads	
W-P2	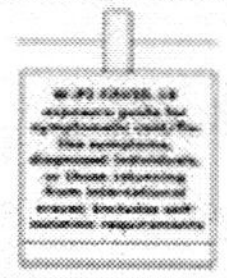	Check-in with 100% of employees & contractors who have self-declared and are isolating	Check-in can be by video, phone, or e-mail. These are people who are in self-isolation due to travel, exhibiting symptoms, or potential exposure to an infected individual.	Operations Manager	
W-M1		Verify temperature screening device is accurate and being used on 100% of all entrants	This check ensures the device being used to record temperatures of entering workers is properly calibrated as per the manufacturer guidelines. It also verifies that the temperatures of all entering workers were taken and those exceeding the upper limit were not permitted entry.	OH&S Lead	
W-M2	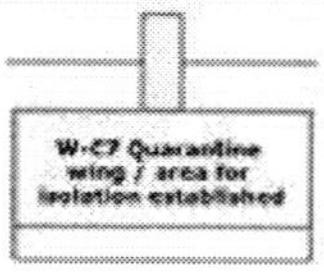	Cleaning schedule in place with 100% compliance	This check applies specifically to areas set aside for site quarantine of people experiencing symptoms of COVID-19. Such workers are restricted to this quarantine location while awaiting transportation from the site.	Facility Management Lead	

Figure 5: Self-Verification (LOD1) Checklist Example

The application of the LOD model to COVID-19 response enables a real-time view of the health of barriers, ensuring all non-conformances and deficiencies are highlighted, together with a risk profile for each facility, office, or entity under consideration.

KPI dashboards tracking compliance with the assurance and verification steps for the more common applications of equipment or activities deemed 'Safety Critical' can adopt a higher level of complexity than would be required for managing risk in a pandemic setting. A relatively simplistic traffic-light approach can serve to easily and effectively highlight how the identified barriers are being adopted across an organization.

A representative example of a KPI dashboard is depicted in Figure 6.

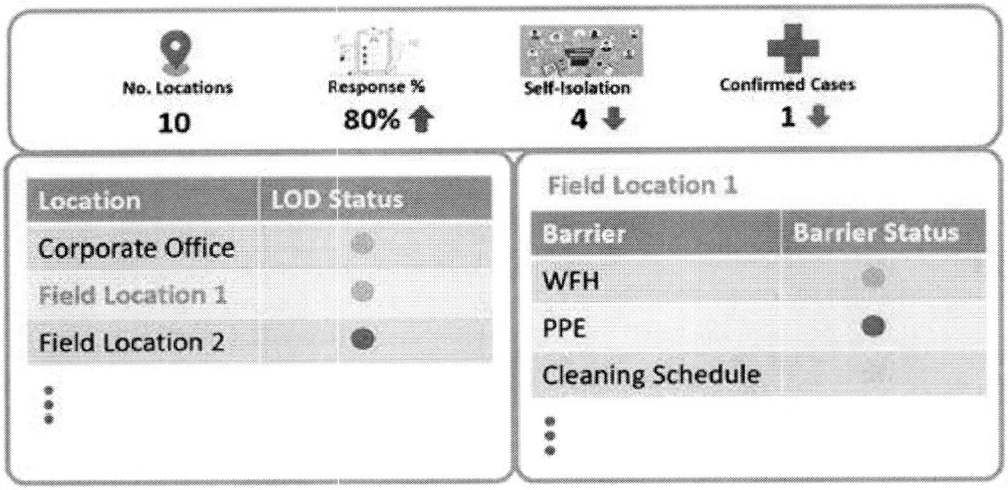

Figure 6: KPI Dashboard Example

4 Conclusion

COVID-19 continues to present a unique and challenging environment for society and industry alike. As organizations continue operations, begin return-to-work preparations, or simply prepare for the everchanging restrictions placed upon them, the need to manage the risks inherent in such activities is paramount.

Bowtie analysis, and its application within an integrated risk management process, can help us better understand and manage these risks by:

- Ensuring that the key hazards have been identified and appropriately assessed; including correct identification of threats, consequences, and both preventive and mitigative barriers;
- Developing, implementing, and tracking a set of assurance and verification tasks, in line with the LOD model, to ensure performance of the identified barriers;
- Highlighting any deficiencies in safeguarding, or non-conformances with the developed assurance and verification tasks, and
- Ultimately providing assurance that the identified risks are being managed / mitigated in a manner which is considered to reduce this risk to an acceptable (or As Low As Reasonably Practicable – ALARP) level.

5 References

1. University of Queensland, Australia, Minerals Industry Safety and Health Centre, National Minerals Industry Safety and Health Risk Assessment Guide.
2. Primrose, M.J., Bentley, P.D., van der Graaf, G.C., Sykes, R.M. Shell International Exploration and Production B.V., "The HSE Management System in Practice-Implementation," SPE 35826, 1996.
3. Primrose, M.J., Bentley, P.D., van der Graaf, G.C. Shell International Exploration and Production B.V., "Thesis – Keeping the Management System "Live" and Reaching the Workforce," SPE 336034, 1996.

4. Gower-Jones, A.D., van der Graaf, G.C., Milne, D.J. Shell International Exploration and Production B.V., "Application of Hazard and Effects Management Tools and Links to the HSE Case," SPE 36031, 1996.
5. Centre for Chemical Process Safety (CCPS), Bow Ties in Risk Management, A Concept Book for Process Safety, 2018.
6. The Institute of Internal Auditors (IIA), IIA Position Paper: The Three Lines of Defense in Effective Risk Management and Control, January 2013.

619345 Relief System Design Revalidation Warranted during Current Pandemic

Abstract:

Numerous operating companies globally are faced with major unparalleled challenges during these disruptive times. Companies are taking radical actions to reduce capital and operational expenditure. One common action taken to limit the spread of the virus is reduction in number of on-site personnel or "only essential to operations" personnel to be on site. As the current wave is looking more ominous compared to what we saw few months back, one can expect these measures getting further harder and stricter. But at the same time, impact of such actions on process safety must be understood. API Standard 521 allows credit for operator intervention and administrative controls to reduce likelihood of overpressure scenarios from occurring. The actions taken during current pandemic can certainly overburden operators jeopardizing administrative controls and compromising overpressure protection. This presentation dives deep into relief system design where administrative control and operator intervention credits have possibly been taken during original design and sheds light on the scenarios needing re-evaluation.

616541 Driving Continuous Process Safety Improvement from Investigated Incidents

Abstract:

Nearly everything we do today to prevent process safety incidents originated in conditions similar to those that led to past incidents. Our engineering forebears recognized the need to learn from past incidents at the beginning of the industrial revolution, and they and we have been building the modern practice of process safety since then. So, after more than 200 years, why do incidents continue to happen? More specifically, why do we as an industry allow incidents with the same root causes to happen repeatedly? Why are we as industry struggling to learn from these incidents?

The Center for Chemical Process Safety (CCPS) recently completed a project to understand the barriers that prevent findings from past accidents from translating to permanent corporate change that prevents future incidents, and develop a process to overcome those barriers. The resulting book, "Driving Continuous Process Safety Improvement from Investigated Incidents," introduces the Recalling Experiences and Learning (REAL) Model. The REAL Model has been designed to help companies move past simply having practitioner awareness of the learnings from incidents to the desired end state where the organization has embedded and institutionalized the learnings within their programs and procedures. The REAL Model ties a company's process safety expertise to specific improvement goals, mines useful information from previous incidents, translates this information into specific improvement and change plans, and describes ways to continually reinforce the reasons for the change so lessons-learned become embedded in the way things are done. Through this process, the learnings become institutionalized within our organization and then future incidents can truly be prevented, we are no longer dependent solely on practitioner knowledge to drive improvements.

This paper will introduce the REAL Model, and present a hypothetical but realistic scenario as an example how a company would use it to drive improvement.

617664 Lessons Learned - How to Make Them Stick

Abstract:

After an incident investigation we gather the facts, assign causes and compile a set of lessons that should, if learned well, prevent a similar incident from occurring. Unfortunately, too often we see the same type of incident happening, sometimes in the same company that had the original incident, or in a company using the same process technology as the company that had the incident. What happened? Are we not learning the lessons? Not because we are not trying. Besides monthly AIChE bulletins, a joint GCPS track on the subject, CSB videos, and countless articles – in the last twenty years 297 articles on lessons learned have been published in *Process Safety Progress* – we continue to have similar incidents. The reason is that we don't learn by listening to a talk or reading a bulletin, at least not for the long term. Our memory fades with time. Even if not, we still need to relate the incident findings to the planning and execution of our work. Thus, the way to remember those lessons is by making them vivid when training, adding them to a well-designed knowledge repository, incorporating them in our technology and daily operations, and periodically reinforcing them. If we manage to do that, there is a good chance that the lessons will be woven into the fabric of the company and incidents will not be repeated. These methods will be detailed in the paper.

617827 Drastic Reduction in Repeat Incident Utilizing Neuroscience Approach (Visual Thinking)

Abstract:

Human are one of key link to repeat incidents. Its observed that usually organizations focus on hardware improvement after an incident but comparative efforts to ensure that associated teams also consistently perform as required and remain alert with associated threats are only limited to training and awareness.

Neuroscience has proven that human brains thinks in pictures or key words i.e. Visual Thinking. This papers will describe how unique approach has been adopted by Fatima site to devised a comprehensive program to utilize this brain limitation in their support. This program not only helped in controlling repeat incidents (mainly related to injuries & Process Safety incidents) but also kept sense of vulnerability intact.

This program not only helped to improve safety performance of site. a site having around 900 contractors and around 800 employees able to avoid any lost time incident has been recorded since over 8 and half years and site is clocked over 57 Million Safe Man Hours.

SPRING21
+17TH GCPS
A Joint AIChE and CCPS Meeting

Writing in the Margins: Considerations for Safe Operating Limits

Sean J. Dee
Exponent, Inc.
4580 Weaver Parkway, Suite 100
Warrenville, IL 60555
sdee@exponent.com

Russell A. Ogle and Matthew S. Walters
Exponent, Inc.
4580 Weaver Parkway, Suite 100
Warrenville, IL 60555
rogle@exponent.com and mwalters@exponent.com

Prepared for Presentation at
American Institute of Chemical Engineers
2021 Spring Meeting and 17th Global Congress on Process Safety
Virtual
April 18 - 22, 2021

Writing in the Margins: Considerations for Safe Operating Limits

Sean J. Dee
Exponent, Inc.
4580 Weaver Parkway, Suite 100
Warrenville, IL 60555
sdee@exponent.com

Russell A. Ogle and Matthew S. Walters
rogle@exponent.com and mwalters@exponent.com

Keywords: enter a few keywords here to aid searching

Abstract

When learning how to write, school age children are often told "Do not write in the margins." This guidance is important because it instills in a child the sense of where text is best placed on a piece of paper. The same is true for chemical processes – normal operating limits define the area where a process is best operated. However, processes may occasionally stray into the margins beyond the operating limits. As a result, facilities may need to define safety margins where temporary operation beyond the normal operating limits is tolerable. In this paper, guidance will be provided for understanding and setting operating limits and safety margins for chemical processes. Good practices from industry for defining and setting limits, as well as documenting and communicating the limits to operational staff will be discussed. Lastly, a case study that examines safe operating limits, alarm set points, interlocks, and process safety information for a dryer processing a thermally unstable material will be presented to illustrate the concepts of the paper.

1 Introduction

Safe operating limits are an important consideration in both understanding and managing the risks associated with chemical processes [1]. Specific definitions for the term "safe operating limits" may vary across organizations and standards, but the definitions typically convey the concept of the range of acceptable values for variables (temperatures, pressure, level, etc.) that fall within the design constraints of a system. These definitions also typically include a consideration of the process dynamics and controls that are in place to address deviations, and as a result, the safe operating limit window may be smaller than the design envelope for a process. A further consideration to note is that safe operating limits may also be a term defined in the context of governmental or legal regulations. As an example, in the United States safe operating limits are discussed within the context of the Occupational Safety and Health Administration's

Process Safety Management Standard (OSHA PSM Standard). These factors (design, process dynamics, and regulations) may all play a role in the definition and development of safe operating limits. The regulatory implications related to safe operating limits will not be discussed here. Rather, this manuscript will focus on the technical information that is necessary to define, understand, determine, evaluate, and communicate safe operating limits in the context of a risk based approach to process safety.

2 What is a Safe Operating Limit?

In *Guidelines for Engineering Design for Process Safety [2]*, "Safe Operating Limit" is defined as:

> *A value for a critical operating parameter that defines the equipment or process unit safe operating envelope beyond which a process will not intentionally be operated due to the risk of imminent catastrophic equipment failure or loss of containment. Operational or mechanical corrective action ceases and immediate predetermined actions are taken at these critical operating parameter values in order to bring equipment and process units to a safe state.*

The Guidelines then provide a graphic representation of the zones of operation, including the safe operating limits as shown below. In the center of the figure is the "Normal Operating Zone." The Normal Operating Zone corresponds to the range defined by the minimum and maximum values of a critical operating parameter that correspond with normal operations. On either side of the Normal Operating Zone are "Troubleshooting Zones." The Troubleshooting Zone is a range where operators can make adjustments to attempt to bring the variable back within the Normal Operating Zone. The Troubleshooting Zone should take into account the response time necessary to diagnose and respond to a deviation from the Normal Operating Zone. Beyond the Troubleshooting Zone is the "Buffer Zone." The Buffer Zone corresponds to a margin in place to ensure that the critical operating parameter does not exceed the safe operating limit, which defines the boundary of the Buffer Zone. Beyond the Buffer Zone is the "Unacceptable / Unknown Operating Zone" which represents an operation state beyond the safe operating limit. In this zone, the process may exceed the instrument range and the equipment's design limit. Process are not intended to be operated in the Buffer Zone or the Unacceptable / Unknown Operating Zone, because as the definition of safe operating limit states, there is a risk of imminent catastrophic equipment failure or loss of containment in this zone.

One of the key concepts of the safe operating limit is that it encompasses *both* the normal operating and troubleshooting zones. Troubleshooting potentially involves the simultaneous recognition, diagnosis, and correction of a process deviation. Once the safe operating limit has been reached, the time for troubleshooting has ended and the focus should shift exclusively to returning the process to the safe operating window. For these reasons, it is common for the regions above the safe operating limits to be characterized as emergency response which may include very clear, direct responses from operations (e.g. activating an emergency system such as cooling/quench systems, relief systems, emergency stop sequence, etc.) or the activation of automated safety instrumented systems [3, 4].

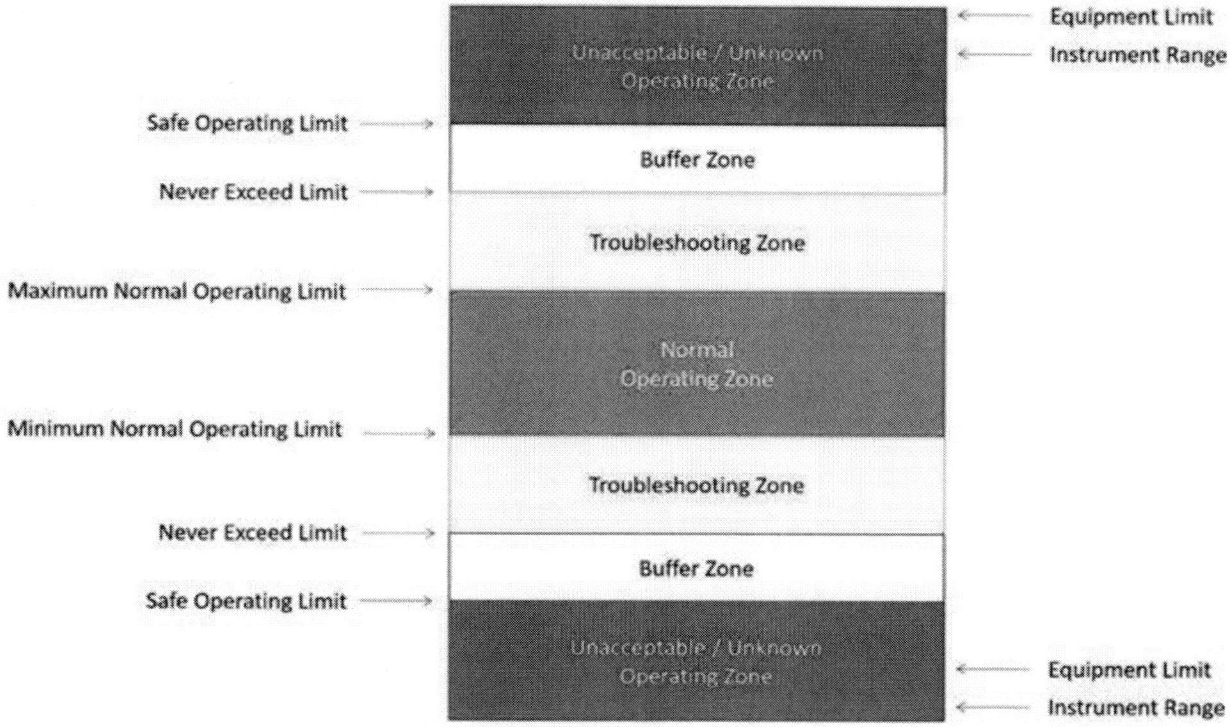

Figure 1. An Illustration of Zones of Operation based on the guidance provided by CCPS [2].

A common critique of safe operating limits literature is the need for consistency in terminology. Recently, the CCPS representation shown in Figure 1 above has been criticized for including the "Never Exceed Limit" below the "Safe Operating Limit." [3, 4]. While the CCPS Guidelines book does not formally define "Never Exceed Limit," it may imply an operational philosophy point that provides ample buffer for activation of emergency response systems to prevent exceeding the safe operating limits (such as a HiHi or LoLo alarm limit). However, others have noted that "Never Exceed Limit" (or similar terminology) can be used in other contexts representing values above the safe operating limits (e.g. the design limit, or closest allowable value to a design limit). One paper provides the example of safety instrumented system design, which uses a never exceed limit to help define trip points, and this limit may exist above the safe operating limit [4, 5]. The example illustrates the importance of understanding the definition of terms and the context in which the terms are used as it relates to safe operating limits. It also underscores the importance of documentation of the various limits and their design bases for a process.

One last consideration is that there may be vast variations in the relative limits for the defined zones of operation. There is not a one-size-fits-all solution for defining normal operating, troubleshooting, buffer, or unacceptable zones of operation. The established limits may vary across different portions of a facility and across different variables. These variations can arise due to differences in equipment, controls, materials, or hazards. However, a relatively consistent methodology can be employed in evaluating and determining safe operating limits, even if the results may vary.

3 How Do You Determine Safe Operating Limits?

There is limited guidance related to good practices for determining the safe operating limits. As an example, the CCPS Guidelines provide three requirements for safe operating limits [1]:

1. Safe operating limits should only be established for critical operating parameters, that have the potential for a catastrophic equipment failure or loss of containment if the limit is exceeded.
2. Safe operating limits should be based on the most constraining design limits of equipment.
3. Safe operating limits should consider the safety and environmental consequences of relief system activation.

As shown in the list above, the primary guidance for defining the safe operating limit is to base the limit on the equipment design limit. In general, this tends to be the extent of guidance in the literature. However, several unanswered questions remain. First, should the safe operating limit be set close to or at the design limit, or closer to the normal operating limit? Second, what should facilities do for equipment that does not have an established "design" limit for particular equipment? Lastly, what if the "design limit" is no longer applicable because a facility has modified its process without properly managing the change? In these circumstances, a facility may rely on a variety of other sources of information to determine safe operating limits. These sources of information could include consulting industry specialists, reviewing operating experience, extrapolating from the normal operating range, employing mathematical modeling, or "nudging" (i.e. gradually changing the safe operating limit under heightened scrutiny to identify any potential adverse consequences) [6].

Few approaches for determining safe operating limits have been discussed in the literature. One group discusses a five step heuristic for setting safe operating limits [7]. The five proposed steps are listed below. It should be noted that this particular group advocates for setting safe operating limits close to the normal operating limits in an attempt to maximize safety risk. The group concedes that this approach (compared to setting the safe operating limit to the equipment design limit) may make assigning consequences of deviation more challenging and complicate the potential expansion of the normal operating limits with addition paperwork associated with management of change.

1. Characterize the operating envelope for all modes of operation
2. Identify key limit values for process parameters
3. Add limit values commensurate with process complexity and risk
4. Drive safe operating limits as close to the operating envelop as reasonably practical
5. Evaluate consequence of deviations

Many of these issues share a striking resemblance to issues with microbial food safety. Food safety from harvest to consumption is accomplished using a Hazard Analysis and Critical Control Point (HACCP) system [8]. HACCP plans involve seven basic principles listed below. The plans aim to identify Critical Control Points (CCPs) for food safety hazards and set the control limits for the CCPs. The plans address corrective actions, as well as the monitoring, verification, and records that are needed to ensure the system is robust.

1. Perform a Hazard Analysis
2. Determine Critical Control Points (CCPs)
3. Set critical limits
4. Establish a monitoring system
5. Establish corrective actions
6. Establish verification procedures
7. Establish recordkeeping procedures

In both lists, the initial focus is to determine the operating limits for the process, determine the subcategory of variables that correspond to critical points that could impact safety, and then establish limiting values. Both methodologies also address corrective actions for process deviations. One benefit of the HACCP list is that it further extends beyond the consequence of deviation and addresses monitoring, verification, and recordkeeping, which can all prove beneficial in terms of maintaining a robust safe operating limit program.

In practice, there will be many competing objectives in the setting of safe operating limits. Facilities may prefer to set safe operating limits as close to the equipment design limits as possible. This provides some degree of operational flexibility and affords opportunity to troubleshoot and correct deviations before the conditions dictate more dramatic emergency response. Alternatively, some may argue that a "margin" be accounted for that sets the safe operating limit below the design limit. However, these "margins" are often arbitrary and lack engineering analysis or supplemental documentation. As previously discussed, some groups also advocate that safe operating limits should be driven as close as reasonably practical to the normal operating limits [7]. This approach maximizes the potential safety margin for the process with a tradeoff of lower operational flexibility.

Where the safe operating limit is set is critically important, but as important (if not more important) is understanding the basis for ***why*** the safe operating limit is set to its prescribed value. Literature (including this manuscript) is typically silent on addressing the question of whether the safe operating limit should lie closer to the design limit versus the normal operating limit. However, one key focus should be on the justification and documentation of the safe operating limit. Often, there are other interplaying factors that need to be considered, whether the design limit or a normal operating limit is used as a primary basis for the safe operating limit. For example, the maximum allowable working pressure (MAWP) is one equipment design limit that commonly discussed in the context of safe operating limits for pressure. The impact of temperature on material strength, and the vessel's failure point (burst pressure) compared to the MAWP may warrant consideration in determining a safe operating limit [6]. Additionally, the safety margins used in code calculations (such as those related to the MAWP) can also change over time [9]. Even with these additional factors, some may still find it reasonable to exceed the

MAWP for brief periods of time, provided that the integrity of the vessel is verified before it is returned to service [6].

Determining safe operating limits can be challenging if there are no explicitly applicable equipment design limits. The safe design and operation of chemical reactions is one example where the safe operating limit may be based more on a normal operating limit, rather than an equipment design limit. In this situation, a process hazard analysis can be used to determine the safe limits to prevent chemical reaction hazards associated with thermal runaway, decomposition, and material scale up [10]. The process hazard analysis consults multiple types of information including literature surveys, experimental and/or computational research, subject matter experts, and potentially regulatory guidelines to understand the safe, acceptable range for operation. After setting the safe operating limits, preventive and protective measures such as alarms and safety instrumented systems can be designed, as well as the basic process control system and operating procedures.

Regardless of whether the preference is to set the safe operating limit closer to the normal operating limit or the design limit, the bases for the safe operating limit is critically important. This methodology and justification of the safe operating limit should be documented to assist with potential management of change analysis that will be required if the value needs to be revisited in the future. Furthermore, the complexity presented by interconnected equipment and the process may make the determination of an appropriate safe operating limit challenging. As a result, more rigorous approaches to determining the safe operating limit may be warranted.

4 Communicating and Promoting Safe Operating Limits

Safe operating limits are a particularly important topic in the United States because they are referenced within the requirements of the OSHA PSM Standard [11]. Specifically, the PSM Standard notes that safe upper and lower operating limits, the consequence of deviating from those limits, and necessary steps to correct or avoid deviation should be included in process safety information, and noted within the operating procedures. As a result, a number of references also address topics related to effectively conveying this information to operational personnel within process safety documentation and operating procedures. The distinction between the safe operating limits and other threshold values that relate to troubleshooting activities (but not the absolute safe limits of a process) may need explicit clarification in the context of a procedure [12]. Approaches for summarizing and conveying the information may vary, but recommendations typically include making the information easy to find and refer to, formatting the information so it is easy to digest (in a table, rather than in a written step of a procedure), and including consequences of deviations and recommended corrective actions in close proximity to the safe operating limits data [13, 14]. While there are many potential methods that can be employed to communicate the information, the focus should be on making the information accurate, technically sound, and easy to understand. In many circumstances, an operator may be referring to this information during a critical stage of response to an alarm or a developing incident. Therefore, streamlining potential decision making based on the information can be beneficial. This also emphasizes the importance of training. Safe operating limits are a critical component of the operational training program, because the operators of the equipment need to know, and remember, the safe operating limits for the system.

Safe operating limits are important only if the management system makes them important. The successful implementation of safe operating limits depends fundamentally on the actions of the organization's operators and workers. The management team must demonstrate their commitment to safe operation and communicate clearly the expectation that safe operating limits are an essential guide to the conduct of operations. It should be noted that operating outside of the safe operating limits can result in multiple risks including safety, quality, and regulatory. Operators and workers must provide their process knowledge and work experience to the formulation of safe operating limits. Process automation, alarms, and data historians certainly play a role in maintaining safe operating limits, but people are at the center of safe operation. Both the management team and the operators must be committed to achieving operation within the safe operating limits. Furthermore, they must be committed to take the appropriate corrective action when operations deviate from those safe limits. This commitment is achieved through organizational discipline.

Operational discipline—the performance of all tasks correctly every time—is another of the foundational elements of safe operating limits. For safe operating limits to be meaningful, operational discipline must be one of the defining characteristics of an organization's culture. If operators sense that the management team will not condone operation outside of the safe operating limits, then operators will be more likely to take safe operating limits seriously. On the other hand, if the management team fails to communicate and reinforce the importance of safe operating limits, then the effectiveness of safe operating limits will be diminished.

5 Case Study – Safe Operating Limit in a Spray Drying System

A facility dried a powder slurry using the spray drying system shown in Figure 2. Liquid slurry was fed to the top of the dryer, where it was sprayed into a mist. Heated air from a natural gas fired air heater was supplied to the top of the dryer. In the main drying chamber, moisture was removed from the sprayed droplets, which dried to a powder that accumulated at the bottom of the dryer and was discharged into a powder conveyor. The exhaust from the dryer was sent to a pulsed baghouse style filter, where fine dust that carried over from the dryer was separated from the exhaust air. The filtered exhaust air was discharged to atmosphere, while the recovered dust was discharged into the powder conveyor.

The facility had three primary temperature measurements in the spray drying system. The temperature of the air heater outlet (T_{heater}) was set to 500°F. The dryer exhaust temperature ($T_{Exhaust}$) normally operated at 250°F, and the baghouse temperature on the clean side of the baghouse filters ($T_{Baghouse}$) normally operated at 220°F. The facility had a safety instrumented system that would shutdown the system if the dryer exhaust temperature ($T_{Exhaust}$) exceeded 300°F, which was the reported safe operating limit for temperature based on the original design.

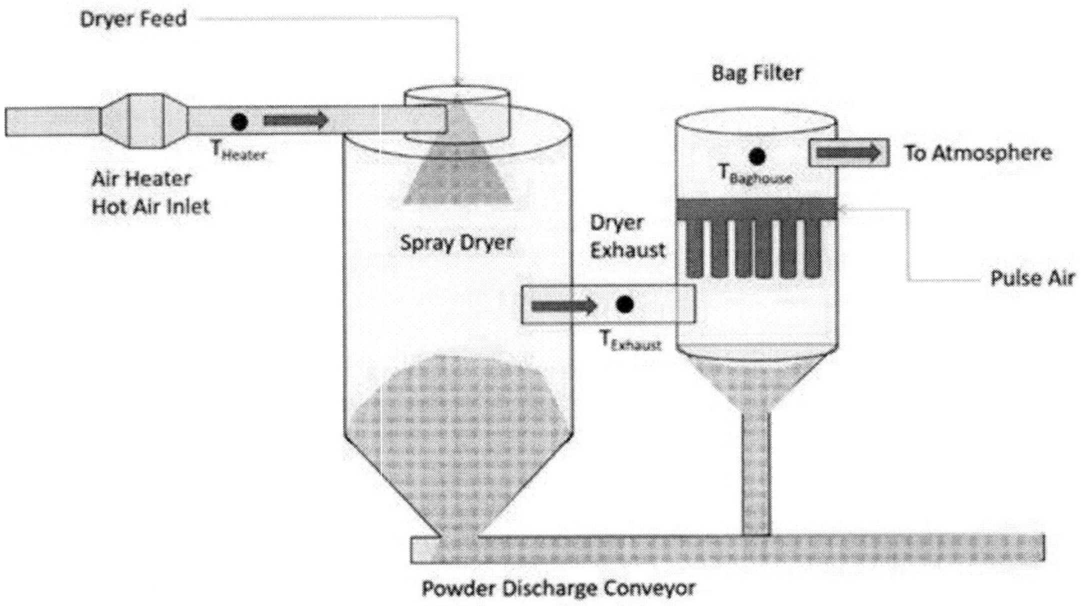

Figure 2. Spray Drying System for Case Study

As a result of an upset in upstream equipment, the facility inadvertently lost feed to the spray dryer. Without the flow of slurry to the dryer, water could not absorb heat from the heated inlet airstream, and the temperature in the dryer increased to 300°F over the course of an hour. The automated safety system activated and shutdown the system without apparent incident. However, as the facility performed its post shutdown checks, a fire was identified in the bottom of the baghouse style filter. The fire alarm was manually activated, and the fire was extinguished. No injuries were reported, but the equipment was severely damaged. Post incident investigation ruled out secondary ignition sources such as hot work, or an ember being transmitted from the dryer to the baghouse. It was clear that the cause of the fire was self-heating of the accumulated powder in the baghouse.

Self-heating in the baghouse was particularly puzzling to the facility for multiple reasons. First, the safe operating limit for temperature in the system (300°F) had been established based on the thermal stability of the product powder. Second, this safe operating limit had not been exceeded during the loss of slurry feed, temperature records showed that the safety instrumented system intervened and the maximum temperature on the dryer exhaust had not exceeded 300°F. Additionally, the baghouse operated at a lower temperature compared to the dryer due to the combination of it being downstream of the dryer, and the input of ambient temperature pulse air into the baghouse filters. As a result, the maximum temperature in the baghouse had not exceeded 270°F. The facility assumed if self-heating were to cause a fire, it should have initiated in the dryer, which was warmer than the baghouse.

The thermal stability of the materials was further evaluated. Historical design documentation included results of accelerating rate calorimetry (ARC) testing that demonstrated the ability of the material to self-heat. Test results reported a 300°F onset temperature for the self-heating exotherm, but three important factors were noted in the review of the original test data.

1. The tested sample product had been reformulated since the original design. The product run during the incident had a slightly different chemical composition, but similar particle size and moisture content as the original test sample.
2. The tested sample was collected from an access port at the bottom of the primary drying chamber. The representative nature of this sample, which would potentially have higher moisture content and larger particle sizes compared to the finer dust in the baghouse, was questioned.
3. The tested sample was approximately 5 grams. The amount of material that accumulated in the bottom of the baghouse and the spray dryer were both significantly larger than the test sample. For these reasons, the appropriateness of using the onset temperature from the test, with no margin to account for thermal stability changes due to scaling, was also questioned.

New samples were collected and subjected to ARC testing. The results for a sample collected from the bottom of the spray dryer were consistent with the original documentation, and showed an onset temperature at approximately 300°F. This confirmed that the reformulation of the product had not significantly altered the thermal stability and tendency to self-heat. A sample collected from the baghouse showed slightly lower thermal stability, with an onset temperature of 295°F. This result was considered consistent with the results of the sample collected from the drying chamber, but demonstrated that the finer, slightly dryer dust in the baghouse could be slightly more sensitive to self-heating hazards. This result (finer dust may have a slightly higher tendency to self-heat) is also consistent with thermal ignition theory regarding particle size and packing density for materials where the self-heating is related to a surface reaction [15].

The ARC test results pointed the investigation to the question that is one of the primary focuses on this paper. What should the safe operating limit be for the system, if experimental data suggests that a hazard manifests near 300°F? How far below 300°F should the safe operating limit be set to ensure that accumulated powder in the system does not self-heat? Again – limited guidance is available, but one reference on determining chemical reactivity hazards points to a common "rule of thumb" of 50°C lower than the corresponding screening temperature [16]. More rigorous testing can be utilized to understand the relationship between size, temperature, and thermal stability. This approach looks at the thermal stability of various geometries, and then develops a correlation between the critical size dimensions and the critical temperature that corresponds to the threshold of thermal instability based on an analysis approach developed by Frank-Kamenetskii [17]. In terms of corrective actions, the following recommendations were provided to the facility.

- Evaluate the impact of setting the safe operating temperature to 250°F for the spray drying system.
    - Consider performing more rigorous testing and Frank-Kamenetskii analysis to determine the critical temperature and critical size relationship for materials handled within the process.
- Develop detection and response plans for locations within the system were material can accumulate and potentially be exposed to temperatures in excess of the safe operating temperature limit.
    - This included considerations of surfaces near the top of the drying chamber where deposits could form, over dry, and self-heat to the point of smoldering.

- Implement a management of change program to evaluate the impact of changes, such as changes to product formulation, on the safe operating limits of the system.
- Evaluate the procedures in place for responding to "hot shutdowns" to ensure that the remediation responses are appropriate if the system has experienced a high temperature excursion.
- Review industry guidance for spray drying system and instrumented controls.
    - As an example, NFPA 61 recently added a new section addressing spray drying systems in the 2020 revision [18].

6 Summary and Key Takeaway

A review of safe operating limits was presented and strategies for selecting safe operating limits was discussed. Some of the key conclusions are listed below:

- Safe operating limits encompass both the normal operating and the troubleshooting regions of operation. As a result, once the safe operating limit has been exceeded focus should shift from troubleshooting to emergency response.
- Relatively limited guidance is provided in literature regarding good practices for determining safe operating limits. While different strategies are apparent, documenting the justification of the safe operating limit is one of the most important aspects of any methodology that is used to set a safe operating limit.
    - Process complexity and interaction between different process variables and equipment can make safe operating limit determination complicated. Caution should be exercised when design limits are adopted as safe operating limits without further analysis of these complicating factors.
- Two separate methodologies were discussed for setting safe operating limits and commonalities were highlighted. In general, good practices for determining or evaluating safe operating limits could include:
    - Perform a hazard analysis to characterize the hazards of the materials handled within the process.
    - Within the hazard analysis, consider the impact of the process dynamics and controls in order to respond to deviations from the normal operating range.
    - Identify reliable sources of information to understand the potential safe operating window of equipment given the materials used in the process. These sources of information may include scientific literature, subject matter experts, experimental or computational research, manufacturer documentation, and/or operational experience.
    - Consider potential failure modes or upset conditions
    - After consideration of the above elements, define the operating range, design limits, and assign safe operating limits between those two values.
    - After safe operating limits have been established, add in alarm and safety instrumented system set points.
    - Operating procedures should clearly state the safe operating limits and the consequences of deviating from safe operating limits. Additionally, operators should be educated on these concepts as part of their training.

- Management systems should be in place to monitor and record system performance. If safe operating limits are exceeded, investigations should determine why the exceedance occurred and then certify that the equipment is fit for service after having exceeded the safe operating limits.

- A case study reviewing the concepts was presented. The case study focused on the determination of safe operating limits in drying systems.
  - Understanding the limitations of experimental data is critical to establishing appropriate safe operating limits.
    - In many instances, temperature safe operating limits are based off thermal stability testing and reported onset temperature of exotherms. However, these temperatures may vary significantly with scaling relationships, and masses that are larger than tested samples may exhibit instability at temperatures lower than the reported onset temperature for a test sample.
  - Process considerations beyond control dynamics and response time may be important in detection of deviation and monitoring adherence to the safe operating limits
    - Operational upsets, such as shutdowns on high temperature, should be considered while determining safe operating limits.
    - Location of sensors may not correspond with the maximum or minimum readings for a particular variable within a system. In these situations, further considerations may be necessary to address setpoints that correspond to upset conditions.
    - Variations in the materials handled within a system may also change testing results. The potential representativeness of testing samples should be verified for safe operating limit determination.
    - Management of change for equipment, process, and material changes can be used to ensure changes do not affect the safe operating window of a process.

7 References

[1] Center for Chemical Process Safety, Guidelines for Risk Based Process Safety, John Wiley & Sons, Inc., New Jersey, 2007.

[2] Center for Chemical Process Safety, Guidelines for Engineering Design for Process Safety, John Wiley & Sons, Inc., New Jersey, 2012.

[3] M. Richardson, Process Safety Management: Standardizing Safe Operating Limit Information, Mary Kay O'Connor Process Safety Center 15th Annual International Symposium, 2012.

[4] T. Stauffer and D. Chastain-Knight, Don't Let Your Safe Operating Limits Leave You S-O-L (Out of Luck), 15th Global Congress on Process Safety, 2019.

[5] ISA–TR84.00.04-2019, Part 1. Guidelines for the Implementation of ANSI/ISA-61511-1:2018, Edition 4 – Revision D, Draft, 2019.

[6] I. Sutton, Process Safe Limits Defining Safe Limits Quantitatively, Available at https://suttonbooks.wordpress.com/article/process-safe-limits-2vu500dgllb4m-4/, Accessed on 2/15/2021.

[7] Inglenook Engineering, Safe Operating Limits – Part 1, Available at https://www.inglenookeng.com/safe-operating-limits-part-1, Accessed on 2/15/2021.

[8] K. Molen, Training Tip: HACCP Plans, Available at https://www.statefoodsafety.com/Resources/Resources/training-tip-haccp-plans#:~:text=The%20purpose%20of%20HACCP%20is,illness%20outbreaks%20will%20be%20avoided, Accessed on 2/15/2021. See figure titled The 7 Steps of HACCP, Available at https://s3-us-west-2.amazonaws.com/abovelms-marketing-assets/Blog+Images/HACCP_HL.pdf, Accessed on 2/15/2021.

[9] W.J. Sperko, Reduction of Design Margin ("Safety Factor") in the ASME Boiler and Pressure Vessel Code in the 1999 Addenda, Available at http://www.sperkoengineering.com/html/articles/Margin.pdf, Accessed on 2/15/2021.

[10] Health and Safety Executive, Designing and Operating Safe Chemical Reaction Processes, Crown, 2000.

[11] United States Department of Labor, Occupational Safety and Health Administration, 1910.119 Process Safety Management of Highly Hazardous Chemicals, 1910.119(d) and 1910.119(f), Available at https://www.osha.gov/laws-regs/regulations/standardnumber/1910/1910.119, Accessed on 2/18/2021.

[12] W.G. Bridges and R. Tew, Best Practices for Writing Operating Procedures and Trouble-Shooting Guides, 13th Global Congress on Process Safety, 2017.

[13] Nevada Division of Environmental Protection, Process Safety Information Program (Revision 3, 2011-03-02), Available at https://ndep.nv.gov/uploads/air-capp-docs/data-ii-psi.pdf, Accessed on 2/18/2021.

[14] T. Jekel, Safe Upper & Lower Limits – Understanding & Documenting, Available at https://www.irc.wisc.edu/export.php?ID=432, Accessed 02/18/2021.

[15] V. Babrauskus, Ignition Handbook - Chapter 9. Self-heating, Fire Science Publishers, 2003.

[16] M.S. Mannan, W.J. Rogers, and A. Aldeeb, A Systematic Approach to Reactive Chemicals, Available at https://www.icheme.org/media/10149/xvi-paper-04.pdf, Accessed on 2/15/2021.

[17] B. Gray, P.J. DiNenno (ed) et al., SFPE Handbook of Fire Protection Engineering (Third Edition) – Section Two, Chapter 10 Spontaneous Combustion and Self-Heating, Society of Fire Protection Engineers, 2002.

[18] NFPA 61, *Standard for the Prevention of Fires and Dust Explosions in Agricultural and Food Processing Facilities, 2020 Edition,* National Fire Protection Association.

618421 Addressing Common PSM Audit Findings on Operating Limits and Safe Limits

Abstract:

Process safety audits are conducted for three main reasons: (1) feedback on process safety program implementation and effectiveness to identify potential improvement opportunities for improved performance, (2) compliance with applicable process safety regulations such as the U.S. OSHA 29 CFR 1910.119 Process Safety Management (PSM) standard, the U.S. EPA 40 CFR 68 Risk Management Program (RMP) rule, and other similar federal, state and/or local regulations, and (3) adherence to any internal company process safety programs/requirements. If a facility has a process covered by the above OSHA/EPA regulations, compliance audits must be conducted every 3 years. Since the OSHA standard was promulgated in 1992, this means that facilities in existence at that time are approaching having conducted 10 compliance audits. Newer processes obviously have had fewer audits, but a review of recent audit findings suggests certain issues continue to be commonly identified. This paper addresses some of the most frequent audit findings observed, specifically related to operating limits and safe limits and the associated consequences of deviations. Guidance is also provided on how these findings can be avoided through appropriate design and implementation of limits tables.

SPRING21
+17TH GCPS
A Joint AIChE and CCPS Meeting

Controlling Human Performance During Unplanned and Planned Tasks within Abnormal Operation Mode

William G. Bridges
Process Improvement Institute Inc. (PII)
1321 Waterside Lane, Knoxville, TN 37922 USA
wbridges@piii.com

PROCESS IMPROVEMENT INSTITUTE

Prepared for Presentation at
American Institute of Chemical Engineers
2021 Spring Meeting and 17th Global Congress on Process Safety
Virtual
April 18 – 22, 2021

Controlling Human Performance During Unplanned and Planned Tasks within Abnormal Operation Mode

William G. Bridges
Process Improvement Institute Inc. (PII)
wbridges@piii.com

Keywords: Operating procedures, SOP, troubleshooting, abnormal mode of operation, non-routine modes of operation

Abstract

There is confusion in terminology used in the chemical-related industry for the class of procedures commonly referred to as Abnormal Mode of Operation and Abnormal Situation Management. This paper provides a clear definition of each mode of operation and gives examples of how the human performance is controlled for each.

1. Normal – Either a continuous mode or a normal batch mode of operation.

2. Planned Non-routine (Non-Normal) – Startup, shutdown, and online maintenance are the main non-routine or non-normal modes of planned operation. But planned temporary procedures, with time limits, are also part of this mode.

3. Abnormal – These include those activities that are planned for and covered by generalized procedures or Trouble-shooting Guides, TSG, but this mode of operation also includes the class of activities that have not been foreseen and therefore do not have written, step-by-step procedures. Companies can and many times do have guides for how to make decisions during abnormal situations.

4. Emergency Operations – These are "planned and partially proceduralized." An emergency operating mode is a diminished or reduced operating plan.

5. Emergency Shutdown – These are "planned and fully proceduralized." Typically, a TSG or Temporary Procedure that fails to resolve the issue in time will go to this; or for some events such as

a sudden loss of containment, the operations team will go straight to one of these procedures.

6. Emergency Response – These are "planned and partially proceduralized" like Modes 4 or 5, but this mode focuses on the protection of people, assets, environment, given the release or other imminent harm is in play.

This paper provides a framework to ensure no mode of operation is overlooked and it will help sites understand what is needed to control risk during each mode of operation. Most of the paper and presentation focuses on Mode 3, Abnormal Mode of operation.

1. All Modes of Operation

There are many modes of operation of a continuous operating process or a batch process. Each mode of operation has unique challenges, but the successful control of each mode depends on controlling human error rates by controlling the 10 core human factors, discussed later in this paper. One good breakdown of the categories of operating modes is provided below:

1. Normal – This is either normal continuous mode of operation or normal batch mode / recipe operation. For a continuous plant, there are no step-by step operating procedures for normal mode, but rather there are general operation orders that dictate the production rates, special operation limitations for the day or week, and a list of triggers (deviations) that point the operators to step-by-step procedures (Abnormal but predicted mode of operation) if the process moves out of the quality bounds or approaches or crosses the safety bounds. Statistics for the past 30 years indicates that about 20% of major process safety accidents occur during normal mode of operation, and about 80% of these (16% of the total number of major process safety accidents) occur due to mechanical integrity failures such as corrosion, erosion, and failure of structural components. A few percentages of major accidents during this mode of operation occurs due to insufficient independent protection layers (IPLs) or inadequate control of the integrity of these IPLs. A few percentages of major accidents attributed to this mode of operation are failures to correctly handle an abnormal situation (see Mode 3 below).

2. Non-routine – These modes are "planned and proceduralized" and include startup, shutdown, and online maintenance as the three main types of non-routine mode of operation; but also planned Temporary procedures, with time limits, are part of these. All these modes have step-by-step instructions (similar to batch operating procedures and the hazards and even the potential accident scenarios can be well understood if the process hazard analysis (PHA) of these modes is performed correctly. Statistics for the past 30 years indicates that about 80% of major process safety accidents occur during these modes of operation. Lack of PHA of procedures for startup, shutdown, and online maintenance is the greatest weakness at most

chemical process plants, because without this PHA on non-routine modes of operation, the process is lacking the unique IPLs needed for these operating modes.

A fourth type of this class of procedure is "Response to failures using a Temporary procedure". Examples include how to run in bypass mode if the flow controller fails and you want to keep running in manual. Temporary procedures are different in that there is normally a "time at risk" consideration for setting a limit on the length of time that a process can be in this mode of operation.

3. Abnormal – Some consider abnormal procedures a sub-class underneath Non-routine operating procedures, but this is not best. Abnormal modes include some un-proceduralized activities, but the industry (and most sites) have guides for most of these abnormal situations, and most of these are triggered from deviations from Normal Procedures (Mode 1 above). Failures during handling of abnormal situations accounts for up to a few percent of the major process safety accidents. A further breakdown of this classification of abnormal mode of operation is possible:

    a. Response to upsets using a Trouble-shooting Guides (TSG) – These are "planned and proceduralized" for handling deviations from the operating window (these or normally triggered on an alarm); these may result in shutdown, safe park, or emergency shutdown (all of which are also proceduralized). One major difference between 3a TSGs and "planned" procedures in Mode 1 and 2 is that the "trouble" predicted in the TSG can occur under many conditions, and the developers rarely put step numbers, but instead organize the TSG into sections and further break the possible actions into "bulleted" steps, with no particular sequence. So, a TSG is planned, but not completely proceduralized as no one can predict whether the process will be ramping up, ramping down, or be in a steady state when the deviation alarm is triggered.

       The trigger for such a response is not always an alarm. The trigger could be a weather report, a flood warning, or some expected issues arising from upsets in supply or upsets in adjacent units.

       Since 1993, the authors have provided several definitive papers on TSGs and how these are developed and how these are documented and administered, the most recent of which is "*Best Practices for Writing Operating Procedures and Trouble-Shooting Guides* [Bridges & Tew, GCPS, 2017]"[6].

       It appears to PII that most of the failures in Mode 3, abnormal mode of operation, are due to weaknesses in brainstorming, developing, writing, practicing, and administering TSGs.

       There are also important limitations of time-constrained responses to TSGs. A key human factors concept reported from investigators at the US Chemical Safety Board (CSB) was humans have a hard time digesting written information. We are pattern-based, not data-based. Mary Douglas wrote that we initially scan about 5 percent of what's in front of us[7]. We match this partial data scan against hundreds and thousands of stored memories, and do a FIRST-fit pattern match, not a BEST-fit pattern match. We "satisfize," we do not optimize. We make decisions very quickly based on partial data recognition. Handheld devices in the field have shown some improvement in this regard, as the information is available a step at a time and as visual information (like a GPS map in your vehicle) can be readily displayed as well.

b. Response to Unanticipated Events – This mode of operation is "not planned and not proceduralized" and therefore this mode of operation tends to worry management the most. In this mode of operation, we rely on experience, knowledge of the process & chemistry, and situational awareness to win the day. The FIRST occurrence of an unanticipated event is handled in PSM through an *Emergency MOC*, which is a way of saying "we'll figure this out immediately, make the best judgment of what to do, do it, and then later we will do the risk review of the change, and then finally we learn from this one event and proceduralize this scenario into a Temporary Procedure or TSG for the next time it comes up".

   The US Nuclear Regulatory Commission (NRC) learned a painful lesson on this Mode of Operation decades ago during Three Mile Island accident, when operators were not given the latitude to make free flowing decisions, but instead were locked into a rigid set of TSGs and emergency response procedures. The aviation industry, governed in the US by the Federal Aviation Administration (FAA) learned about this mode of operation many decades ago, as the number of accidents due to Mode 1, Mode 2, and Mode 3a dropped dramatically, leaving only Mode 3b to deal with. The US NRC and US FAA are now focusing on helping operators (crews) recognize when they are in this mode and on providing the skills and thinking processes to handle this mode of operation.

4. Emergency Operations – These are "planned and partially proceduralized." In summary, an emergency operating mode is a diminished or reduced operating plan, and is normally controlled (at least in part) by a Temporary procedures.

5. Emergency Shutdown – These are "planned and fully proceduralized." Typically, a TSG or Temporary Procedure that fails to resolve the issue in time will go to this; or for some events such as a sudden loss of containment, the operations team will go straight to one of these procedures.

6. Emergency Response – These are "planned and partially proceduralized" similar to Modes 4 or 5, but this mode focuses on the protection of people, assets, environment, given the release or other imminent harm is in play.

2. Coverage of All Modes of Operation within US OSHA PSM and US EPA RMP regulations

The US OSHA regulation 29 CFR 1910.119 on process safety management (PSM) does mention most of the modes of operation listed above, and states that covered facilities should have procedures for each mode of operation, routine and non-routine. But the OSHA PSM regulation does not mention 3b mode of operation, *Response to Unanticipated Events*. Like the US NRC and US FAA before them, the process safety regulators (US OSHA and US Environmental Protection Agency [EPA]) initially assumed that all events could be predicted and therefore "planned for and proceduralized". But this is not completely possible, as the NRC and FAA found out.

Many in the chemical and related process industry recognize that chemical process are at least as complex as nuclear power plants and flying a passenger jet, and so there will be some events or situations that cannot be predicted. There are industry best practices on 3b mode of operation, some more than 40 years old, and there is a growing discipline for handling mode 3b events. And there is a depth of knowledge and experience in handling the other modes of "planned and proceduralized" 6 classes of activities of many decades, though best practices for developing and using TSGs are less than 3 decades old.

3. Coverage of All Modes of Operation within Process Safety Best Practices

Process safety was practiced in the chemical industry for decades before there were regulations on PSM. In fact, the US regulations on PSM are an outline of what the industry was already doing to control process safety, though candidly the US PSM regulations are only a skeleton of the best practices for controlling process safety; and this skeleton is missing many bones!

Many companies do a reasonable job of addressing risk during modes 1-6; and some do a great job. ***But many are still weak on Mode 3a, and most are weak on Mode 3b***. At the best companies, the goal is to shrink 3b types by converting them into 3a (as more of the 3b scenarios are learned), and also by eliminating many of these scenarios by targeted IPLs. But it is highly unlikely industry will learn to predict all events and so a TSG or Temporary Procedure will not be available and practiced (drilled) for some events; those remaining scenarios without procedures are **Mode 3b**.

As mentioned earlier, from statistics from the industry and from US OSHA, most major process accidents occur during Mode 2 and Mode 1. Figure 1 on the next page illustrates this distribution of when accidents happen. The entire category of Other is less than 5% of the number of major process safety accidents. There is not enough data to make a firm conclusion, but it is believed that in the chemical-related industry, less than 1% of the accidents are related to Mode 3b scenarios (unplanned and proceduralized response to an event). This may appear to be good news, but part of the reason the percentage is so small is because the number of losses due to the other modes of operation are all too high compared to the aviation industry and nuclear power industry. For the aviation industry, the fraction attributable to this 3b category is likely 20% or higher.

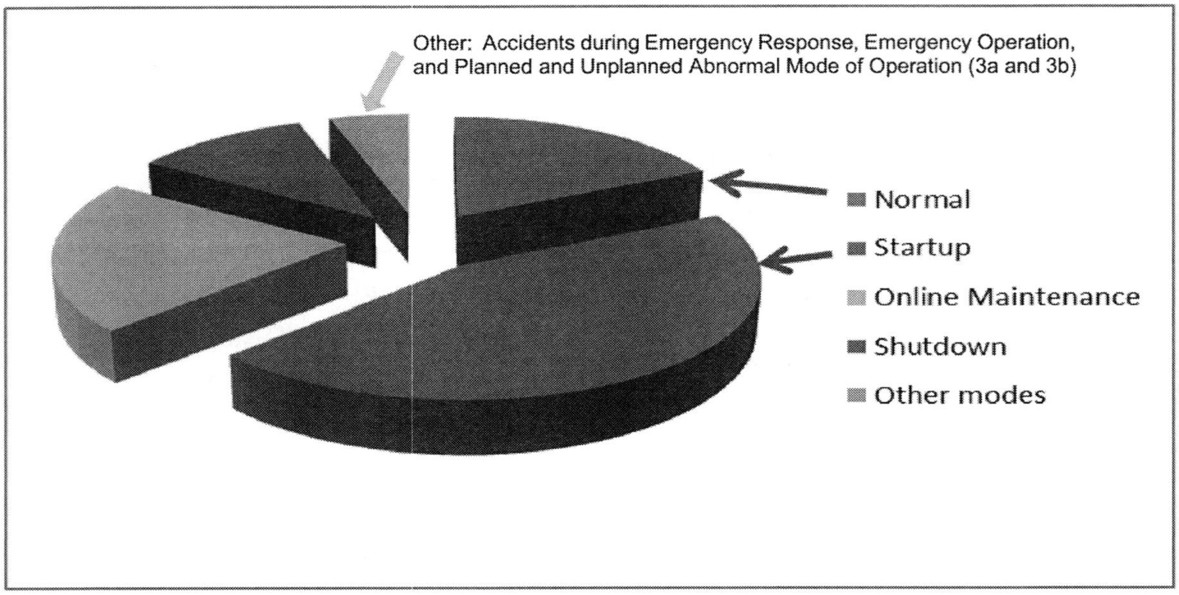

Figure 1: Distribution by Operating Mode of Major Process Safety Accidents[20]

Most of Mode 3 activities arise during deviations and failures during Mode 1 (normal operating mode) which directly take us to a Mode 3a or 3b activity (or directly to a Mode 4 or 5 activity). The best sites can and normally in fact do a great job of developing Mode 3a procedures with the associated plans and drill. Unfortunately, it appears most sites are not doing a good job in controlling Mode 2, 3b, or 4 activities. This is not because of (only) a lack of regulation… this is because of many factors.

PII (and likely others) provide courses on identifying procedures and developing procedures for Mode 2, Mode 3a, and Modes 4, 5, 6. And in PII's management of change (MOC) courses, how to plan for Emergency-MOC (Mode 3b activities) is covered. All of industry struggles with Mode 3b modes of operation. But fortunately, if Modes 2, 3a, 4, 5, are 6 are clearly planned and if there are sufficient IPLs in place for deviations from these procedures, which are predicted during PHA of the deviations from these procedures (which 80% of the industry is ***NOT*** doing), then the remaining number of Mode 3b scenarios becomes pretty small; about 1% of the major process safety accidents.

The more we learn and brainstorm and predict, the more Mode 2 and 3a procedures we will have, and the more the number of residual 3b scenarios (those that are unplanned and un-proceduralized) will shrink.

For Mode 3a scenarios, as mentioned earlier, the paper *Best Practices for Writing Operating Procedures and Trouble-Shooting Guides* [Bridges & Tew, GCPS, 2017][6] provides excellent guidance on planning for and writing the TSG for each predicted deviation. And further, the paper *LOPA and Human Reliability – Human Errors and Human IPLs (Updated)* [Bridges & Clark, GCPS, 2011][4] and the follow-on paper *Proven*

Approaches to Ensuring Operators Can Respond to Critical Process Deviations in Time (Human Response IPL) [Bridges, GCPS, 2017][5] show how the best companies implement effective systems to control accidents during Mode 3b. These three (3) latest papers build upon papers from 1995 until now on how to develop TSGs and approaches to get operators trained and drilled on responding to critical deviation alarms, and when to fallback to Mode 4 & 5 activities.

The next three pages provide examples of the procedure styles associated with Mode 2, 4, 5, 6 operations (Figure 2) and Mode 3a (TSGs) (Figure 3). Mode 1 does not have procedures but rather just has "shift orders" and "operator rounds and logs".

Section 5 of this paper provides focus on Mode 3b. But first, Section 4 will recap some key points related to human error probability for the other modes of operation. Again, the reader should review Appendix A for a listing of all human factors and multiplication factors in error rates, if the human factor is not optimized.

4. Aside: Human Error Probability for a Single Execution of a Rule-Based Task *(the Tasks is Planned and Proceduralized)*

To calculate the probability of human error (P_{HUMi}), the type of tasks must be defined and the baseline error rate for such a task needs to be established. Note that with excellent control of all of the human factors, a company can begin to approach the lower limits that have been observed for human error, but individual, specific human error probabilities may average about $P_{HUMi} = 0.01$. In some applications, such as maintaining SIL 2 and SIL 3 SIFs, $P_{HUMi} = 0.01$ is a large probability and it is critical in such applications to provide detection and correction for specific human errors.

Excellent control of all human factors means a robust design and implementation of management systems for each human factor are achieved with a high level of operational discipline. The first well-researched publication detailing potential lower limits of human error probability was by Alan Swain and H Guttmann (NUREG-1278, 1983)[8] and by others. However, many times, the limits they referenced get used out of context. The lower limits in NUREG-1278 assume excellent control of human factors, but such excellent control is rarely, if ever achieved. Additionally, some human errors listed by Swain and others were for a single error under highly controlled conditions, or on a "best day" instead of average error probability or rate over an average year of tasks. In general, PII has found it best to use the average error probabilities as discussed in the following section.

Figure 2: Example of Procedure that Follows Best Practice Rules for Operating Modes 2, 4, 5, 6

Unloading Monomer from Tanker to Storage

STEPS	DETAILS
1. Wear standard PPE, plus rubber gloves and full-face organic respirator.	Standard PPE includes hard hat, safety glasses with side shields, and steel-toe shoes.
2. Weigh in the tanker.	Record weight on GROSS line of form.
3. Check bill of lading.	... to verify correct type of material is in tanker.
4. Sign in tanker driver.	Driver must sign in as a visitor and be escorted at all times.
5. Take Certificate of Analysis to QC.	***CAUTION: To avoid contamination, DO NOT unload until you receive approval from QC.***
6. Spot tanker.	A) Direct driver to location. B) Verify brake is set. C) Chock the wheels on at least one side between the two rear axles. D) Ground the tanker (attach grounding strap to an unpainted metal surface).
7. Have the storage area operator make sure the storage tank can hold contents of tanker.	*WARNING: Failure to perform this step can result in an overflow and spill of hazardous Monomer.* Perform **Tank Gauging Procedure (SOP-01-804)** or check the storage tank load cell readout. Storage tank operating limits are stenciled on the control panel.
8. Place 5-psig nitrogen pad on tanker.	A) Connect from local 5-psig nitrogen drop to aft bottom nitrogen connection on tanker. B) Open nitrogen supply valve. C) Then open valve at tanker.
9. Connect Monomer unloading hoses.	A) Verify hoses and gaskets are in good condition. (Hoses are normally stored on hose rack at unloading spot.) B) Connect from tanker to air pump. (Use aft bottom connection on tanker. Remove cap.) C) Connect from air pump to storage tank. (Connect to storage tank line labeled "from TW.")

Figure 3: Example of Trouble-Shooting Guide that Follows Best Practices for Operating Mode 3a

Trouble-Shooting Guide

Alarm or Indicator:	**PAL 4446 – Low Pressure Alarm for Suction of Organic Feed Pump 40-PM-18.445**

Action Limit:	5 kPa		
Consequence:	Possible pump seal failure, releasing or spraying organic waste into the berm.		
Process Area:	FB&D Incinerator; Liquid Organic Liquid Feed	**Oper. Mode:**	Normal
Drawing #s:	D-400-PI-013		

IMMEDIATE ACTION (by system or by operator)

- DCS should shut down the organic feed pump (40-PM-18.445).
- From the DCS display, MAKE SURE the organic feed pump is shutdown.
- HAVE the field operator check for leaks near the organic feed pump.
- IF there is a large leak/release, THEN use the ESD switch to shutdown the unit and then follow/complete the shutdown and isolation procedure, OPS-ESD-117.
- IF there is a minor leak or no leak, THEN:
    - COMPLETE the rest of the trouble-shooting,
    - and DECIDE how to contain the leak for now,

DECIDE IF ALARM is REAL

- From the DCS, CHECK the pressure and feed tank level trends. IF the trends indicate the alarm if valid, THEN continue with finding the cause or fixing or bypassing the problem.

FINDING and FIXING the CAUSE

- CHECK valves upstream of the organic feed pump to see if any are closed too far, including checking ESD valves.
- CHECK, by feel with hand, if the heat tracing is on; IF Not, then TURN ON or open heat trace valves
- MAKE SURE nitrogen to the pump seal is at the normal operating pressure.
- CHECK if the line if plugged or frozen (skill)

ABC Chemical Company Prepared by: *Printed copy of this procedure is good for one job task duration.*	OPS-76-TSG-233	Incinerator Unit Revised 8/24/2015 Printed 2/21/16 Page 1 of 2

Figure 3: Example of Trouble-Shooting Guide that Follows Best Practices (continued)

Trouble-Shooting Guide

Alarm or Indicator:	**PAL 4446 – Low Pressure Alarm for Suction of Organic Feed Pump 40-PM-18.445**

- CHECK if the line if plugged or frozen (skill)
- CHECK if the level is actually low, use the Organic Feed Tank.
- IF the cause is low level in the feed tank, THEN resolve the problem if necessary based on cause that is found (skill)

FIX or BYPASS PROBLEM

- IF necessary, SHUT DOWN the Unit to allow fixing of the problem.
- FOLLOW the proper procedure to resolve problems (repair procedure, line clearing procedure, etc.)
- IF the decision is made to continuing operation without all equipment in normal condition, THEN:
    - FOLLOW Temporary Operating mode, if there is a temporary procedure already written for this possible problem/condition
    - FOLLOW MOC procedures to obtain approval for any non-standard temporary operating procedure or mode

END

Error Probability for Rule-Based Actions that are Not Time Dependent (Applies for Modes of operation 1 and 2 only):

Actions that do not have to be accomplished in a specific time frame to be effective are not time dependent; this is the case with Mode 1 and Mode 2 of operation. It should be obvious then that these do not include response to alarms, or similar actions with time limits (Mode 3 are highly time-dependent). Values listed below represent the lower limits for human error rates, assuming excellent control of human factors; these are expressed as the probability of making a mistake on any step:

- 1/100 – process industry; routine tasks performed 1/week to 1/day. This rate assumes excellent control of all human factors. Most places PII visits, the workers and managers and engineers believe this is achievable, but not yet achieved. {Actual data from the Savanah River Site indicated a Miscalibration error probability of 7.0E-3 and a Failure to Restore After Maintenance error probability of 5.1E-3 for an organization with excellent human factors control and data gathering (Table 1[9]). It is noted that these values could be rounded to 1/100. Organizations are cautioned to determine the actual data for human error rates in their own management systems before using human probabilities lower than 1/100.}
- 1/200 – pilots in the airline industry; routine tasks performed multiple times a day with excellent control of human factors. This average has been measured by a few clients in the airline industry, but for obvious reasons they do not like to report this statistic.
- 1/1000 – for a reflex (hard-wired) action, such as either proactive or minor corrective actions while driving a car, or very selective actions each day where your job depends on getting it right each time and where there are error recovery paths (such as clear visual cues) to correct the mistake. *This is about the rate of running a stop sign or stop light, given no one is in front of you at the intersection; the trouble is measuring this error rate, since you would have to recognize (after the fact) that you made the mistake.*

See Bridges and Collazo (GCPS, 2012)[10] for more details on this topic. Also see Appendix A.

Adjusting the lower limit rates to estimate a baseline rate at a site.

As mentioned earlier, the lower limit rates assume excellent control of human factors in the industry mentioned. Note that airline pilots have a lower error rate than what PII has measured in the process industry. This is due, in part, to the much tighter control by the airlines and regulators on factors such as fitness-for-duty (control of fatigue, control of substance abuse, etc.). Excellent control of human factors is not achieved in many organizations; therefore, the human error rates will be higher than the lower limit, perhaps much as much as 20 times higher. Table A1 in Appendix A provides adjustment factors for each human factor. These factors can be used to adjust the lower limit of error rate upward or downward as applicable, but the factors should not be applied independently. For instance, even in the worst situations, we have not seen an error rate for an initiating event or initial maintenance error higher than 1/5, although subsequent steps, given an initial error, can have an error rate approaching 1 due to coupling or dependency.

- 1/5 – highest error rates with poor control of human factors; this high rate is typically due to high fatigue or some other physiological or psychological stress (or combination). This is the upper limit of error rates observed with poor human factors and within the process industry. *The error rates in the Isomerization Unit the day of the accident at BP Texas City Refinery[11] were about this rate. The operators, maintenance staff and supervisors had been working about 30 days straight (no day off) on 12-hour shifts.*

For most applications, assume ***a baseline error rate of 0.02 (1/50) errors per step***, which is about average at the sites PII visited in the past 15 years. This value could be justified based on the fact that most chemical process sites do not control overtime during turnarounds and/or do not have a system for controlling verbal communication using radios and phones. In addition, for critical steps such as re-opening and car-sealing the block valves under a relief valve after the relief valve is returned from maintenance, the error probability is about 0.01 (1/100) to 0.04 (1/15)[12]; plus, the average probability of being in a "fail to function" state at time zero for a relief device is between 0.01 (1/100) and 0.02 (1/50) (Bukowski, 2007-2009)[13, 14, 15]. Both of these tasks have multiple checks and have procedures (similar to what is done when servicing a SIF and when using bypasses for an SIF) and yet the observed human error probability remains between 0.01 and 0.02.

Error Probability for Response Actions that ARE Time Dependent (Applies for Operating Mode 3a) and that are Planned and Proceduralized.

This is the probability that the correct action will be completed within the time necessary, but after that time, the action will not help (so would be the same result as not doing the task). The probabilities listed below assume that there is a written TSG or similar procedure available to guide the action, and that there is a clear and unambiguous call for action, such as a loud alarm and light.

- 1 to 1/10 - if practiced/drilled once per year and there is not sufficient time to accomplish the response task. *It may get done in time on selected occasions, but you cannot count on there always being sufficient time available.*

- 1/10 - if practiced/drilled once per year and if there is always sufficient time, theoretically, to accomplish the response task. *This error rate assumes excellent human factors related to response actions.*

- 1/100 - if practiced/drilled once per week and there is sufficient time to accomplish response task. *This error rate assumes excellent human factors related to response actions.*

Standard rules (from LOPA) for Human Response to Critical Alarms

Typical rules for allowing credit for a human response to an alarm as an IPL, state that the alarm response be unique, independent, dependable, and auditable. More specifically:

- There must be a clear, unambiguous signal that an abnormal condition is present.
- The alarm cannot be generated from a component of the initiating event failure (i.e., credit cannot be taken for an alarm if failure of the measurement is a part of the initiating event)
- Distinct written procedures outlining required operator response to the alarm must be in place; these are normally in the format of a TSG.
- The response must be in the operator certification training program.
- The certification must have repetition, such as drill once a year, with measurement and recording of error rate and success rate of operators accomplishing the steps of the TSG within about ½ of the overall process safety time that is allocated to the "actual response to the alarm" response.
- There must be sufficient time (5 minutes for full response from the control room or 20 to 40 minutes if responding in the field/plant) to intervene after the alarm and before the consequence is manifested.

This level of rigor had a side-effect of minimizing the implementation of Human Response IPLs. Some believed this was a good thing, as their companies want to try to limit Human

Response IPLs to instances where an automated SIF was impractical or not possible. However, if implemented correctly, Human Response IPLs can be more effective than many other types of IPLs.

5. Controlling Human Error During Surprise Events – those situations we have Not Planned for and for which we do not have written procedures or guides

As a reminder Mode 3b is responding to unexpected events without TSGs or procedures. Currently, this mode of operation is commonly covered in the Emergency-MOC program. This might not be obvious to all process safety practitioners, but for a great many of us "emergency-MOC" is code for "we plan to work outside of the TSGs and other procedures because we are in uncharted territory… at least for now" … that is what most folks in the industry call Emergency-MOC. The risk review that is required for each MOC is performed after the change rather than before. All of us strive to eliminate the need for entering operating Mode 3b or we say we "strive to eliminate the need for emergency MOC," but no-one in any industry has accomplished this goal yet. Industry and regulators need to acknowledge that despite our best efforts, we simply CANNOT predict, plan, and proceduralize every possible event.

> ***Real World Example:*** Following the disaster at the Philips polyethylene plants in Pasadena, TX, in 1989, Phillips entered into a settlement agreement with US OSHA to follow all best practices in process safety, including all of the planned requirements in the OSHA PSM standard (still two years away from issuing as a regulation). To be proactive in a meeting with OSHA staff and plant staff, Phillips legal staff announced Emergency-MOCs would no longer be allowed at Phillips operations (meaning, the company will force a shutdown for every situation that has not been predicted and proceduralized, so as not to either have an involuntary MOC or try to make a better decision in the heat of the moment). Several attendees with many years of operations and process safety experience pushed back and stated that there will be a shutdown every few days to a week if this policy of no "emergency MOC" is implemented. The attorneys for Phillips insisted on the new policy. This is a real-intended stance. However, within 2 months, it was reversed and instead the company started to focus on eliminating as many "unplanned events" as possible by doing greatly improved PHAs of all modes of operation while proceduralizing (and training and drilling on) proper responses to all known deviations from the norm.

Key issues with Mode 3b situations: Because Mode 1, 2, 3a, 4, 5, and 6 operational situations are predicted/expected, they are proceduralized. Even in the case of a TSG — where the exact path and response is variable, based on information gleaned by the operator — the various paths have been thought through beforehand and documented. Thus, the responses to these tasks are developed by subject matter experts. There may be many different responses, and we use our best judgment to select one and use it. In other words, cause and effect are known, subject matter experts (SME) provide options, and then we pick a good practice. When the need arises, we SENSE the need; Experts ANALYZE; the worker selects the best option and RESPONDs.

In the words of Steven Cutchen[23], each of these modes are Complicated in that they are not obvious but can be analyzed in a straight-forward fashion to establish the proper course of action.

Real World Examples:

- Making a cheesecake is Complicated. One SME may say use a box mix. Your grandmother, certainly an SME, may have her scratch recipe, and show you how the crust is supposed to feel when the moisture is correct.

- A debottleneck project is Complicated. Different contractors may propose different processes. You pick one, maybe not the cheapest.

In contrast to the above modes, operating Mode 3b situations are unanticipated and they are not proceduralized because they are unpredicted. Cause and effect are not discernible ahead of time and it may not even be possible to create a procedure after such a situation. This makes 3b situations extra Stressful and in the words of Steven Cutchen[23], Complex (i.e., situations in which we cannot accurately predict the outcome by traditional means).

Approaches to Handling Mode 3b Situation:

1. First, the company/site needs to optimize the human factors that govern all modes of operation, include 3b Mode. See Appendix A and the related papers on human factors at this conference[16, 17] and prior conferences[10, 18, 19]. This ensures that that contributing factors such as worker fatigue and verbal miscommunication on radios and phones will not exacerbate the already Complex and Stressful situation.

2. Teach your staff that the organization has made great efforts to predict all situations and have written plans in place for abnormal situations, and that any deviation from pre-planned approaches should follow the established MOC system, which ensures the hazards for a change or new situation is analyzed well before making the change. HOWEVER, also admit and stress that an unknown situation can occur and when it does, the workers (or selected staff) have the authority to:

   a. SHUT DOWN the process

   b. SAFE PARK the process, … or if deemed safer,

   c. CONTINUE TO OPERATE TEMPORARILY (emergency MOC territory) in a mode or method that is currently not documented.

   The site must clearly designate who has this authority for decision making during 3b Modes of operation.

3. Set Constraints and General Guidelines on when to force a shutdown to make that decision more comfortable in an unexpected, unpredicted situation. Note that it may seem best to immediately shut down anytime the process deviates into the unknown. But what if the staff believe a safe shutdown cannot be achieved? Or that following the approved shutdown procedures will not function well in this situation? Or that shutting down is less safe than switching to a unique way to operate temporarily? In such situations, taking away the flexibility to exercise initiative may cause more harm.

4. Brainstorm with those staff at the highest expertise level how to **experiment, test,** or **probe** for ways out of the situation, while maintaining options to fully and immediately shut the process or kill the reaction (last fallbacks). This can and should happen ahead of time using scenarios that had been unknown until a surprise happened. This can also be used in real time for a new unknown event. With each **experiment**, we **sense or perceive** what is working and what is not. Then we **respond** based on what we have learned and prepare for the next step or next probe or finally exit to a known procedure (exit the 3b mode of operation). Tabletop drills are similar in nature, but with those we normally have a known scenario to deal with. But the more times we exercise our minds on how to deal with the unknown situations, the better our real-time "probe and learn" becomes.

During the true 3b scenario, the problem is changing, and the solution is emerging. We rely on our experiences and on our education and knowledge. To be most successful, we rely on the experiences of others as well. We collaborate. Staff can learn to approach new problems, using these steps.

- We troubleshoot (if within the bounds of a TSG; a 3a situation)
- When the need arises (3b situation), we EXPERIMENT within constraints and guidelines. We then SENSE the response to develop our learning. Finally, we RESPOND with our learning to inform the next PROBE.

Real World Example:

- Driving through East Tennessee you will go through many different terrains. The headlights reach 200 feet or so ahead. The precise circumstances of the drive make a procedure unworkable. Yet, the drive is made without difficulty. This is because of experience and the constant practice of making changes and sensing the response and learning what works and doesn't work.

Real World Example Where Initiative was Used Out of Order:

- A senior operator at a petrochemical plant in Saudi Arabia using newly developed chemistry for unique products ended up in a runaway reaction situation. The runaway was both unsafe (more unsafe than the operator likely realized) and a shutdown following the procedure was not quick enough and yet a trip of the reactor using the pre-design "big red button trip" would lead to days of cleanout of polymer buildup in the reactor. The worker did not probe in this case, but instead forced a trip of the hydrogen feed control valve by putting a related controller in a mode that fooled one of the reactor isolation valves to close out-of-sequence (the operator did not have a remote close button for that specific isolation valve), which then cascaded to close all other isolation valves, but in a different sequence than designed. This stopped the unsafe runaway reaction while also preventing nearly all of the polymer build that would have otherwise resulted. This is technically not a 3b event, as the operations staff had "brainstormed and pre-planned" this approach to shutting the reactor down in a "better" way. The main issue in this example is that management and engineering staff were not receptive enough to New Ideas or possible improvements to the design unless such ideas came from a PHA team or Investigation team. So, in preparing for 3b events, each operation should work hard to listen to the plant staff and their ideas and to brainstorm with them on "better ways to make cheesecake." This will reduce the number of 3b scenarios and allow the company to make such improvements ahead of time.

What's Next:

Once a 3b event has occurred and you have learned the best approach, by all means try to proceduralize this in a step-by-step fashion, or in a guideline (TSG) fashion. If necessary and possible, develop a hardware or instrumented solution to the problem (as done in the last example above). The goal at most companies is to shrink the number of 3b types by converting them into 3a or 2 modes. For many Complex problems, having seen the event, it is now known and can be accommodated - designed around and proceduralized. But this can inadvertently create another problem. Other complex situations of 3b nature are inevitably on the way. Many of the remedy tasks cannot be proceduralized even after the fact. So, the structured approach listed above is necessary in most processes. And this concept, along with others, are missing from US PSM regulations and are not documented well in current guidelines from AIChE, API, AFPM, etc.

Note that Steve Cutchen, formerly of the US CSB, presented an excellent paper at the 16th GCPS, online, in August 2020, that described an approach labeled Safety-II[21]; this approach was started in Europe and is finding some acceptance there. This approach is like the one outlined above, and the author borrowed some of the phrasing for some of the points above from e-mail correspondence with Mr. Cutchen[7]. The reader is encouraged to research the Safety-II approach further.

6. Conclusions

The industry understands and controls the risk of accidents during all modes of operation; with a range of success from excellent control to not-so-good. One mode of operation has not gotten much attention – how to handle unplanned situations (unanticipated and un-proceduralized mode of operation). Most abnormal situations are "anticipated" and therefore planned for with hardware remedies or procedure-based remedies. If a site does an excellent job of predicting accident scenarios and controlling these, then what is left is less than 1% of major process safety accidents occurring from totally unforeseen causes.

This is good, since the chemical-related industry has not developed a standardized approach on how to handle the scenarios "they don't know about." Merely labeling them as black swan events and throwing our hands up when they occur is not best (after all, black swans are not really that rare in nature).

This paper laid out the background to understand what can be predicted and proceduralized, and also provides a framework for what to do with the scenarios we have not predicted.

We KNOW there will be unexpected and even unpredicted scenarios… that is but the first step in not being caught completely off guard. For these events, we need to develop "approaches for recognizing and handling" the unknown scenarios in a rational way. The industry needs more cohesion on such approaches.

7. References

1. Bridges, W., and Williams, T., *Risk Acceptance Criteria and Risk Judgment Tools Applied Worldwide within a Chemical Company*, International Conference and Workshop on Risk Analysis in Process Safety, October 21–24, 1997, Atlanta, GA, pp. 13–28. American Institute of Chemical Engineers. New York, NY. 1997.

2. Bridges, W., and Williams, T., Risk Acceptance Criteria and Risk Judgment Tools Applied Worldwide within a Chemical Company, International Conference and Workshop on Risk Analysis in Process Safety, October 21–24, 1997, Atlanta, GA, pp. 13–28. American Institute of Chemical Engineers. New York, NY. 1997.

3. Dowell, A., Layer of Protection Analysis: A New PHA Tool, After HAZOP, Before Fault Tree, International Conference and Workshop on Risk Analysis in Process Safety, October 21–24, 1997, Atlanta, GA, pp. 13–28. American Institute of Chemical Engineers. New York, NY. 1997.

4. Bridges, W., and Clark, T.,, *LOPA and Human Reliability – Human Errors and Human IPLs (Updated)*, 7th GCPS, AIChE, March 2011.

5. Bridges, W., *Proven Approaches to Ensuring Operators Can Respond to Critical Process Deviations in Time (Human Response IPL)*, 13th GCPS, AIChE, March 2017.

6. Bridges, W., and Tew, R., Best Practices for Writing Operating Procedures and Trouble-shooting Guides, 13th GCPS, AIChE, March 2017.

7. Cutchen, Steve "Re: I'll be making two presentations at the 2020 GCPS Virtual Conference - August 20." Message to William Bridges. 28 August 2020. E-Mail.

8. Swain, A. D., Guttmann, H. E., *The Human Reliability Handbook - with Emphasis on Nuclear Power Plant Applications*; NUREG/CR-1278, U.S. Nuclear Regulatory Commission, Division of Facility Operations, 1983.

9. Savanah River PSV failure rate data *(see Reference 13, 14, 15)*

10. Bridges, W., and Collazo-Ramos, G., *Human Factors and their Optimization*, 8th GCPS, AIChE, April 2012

11. US Chemical Safety and Hazard Investigation Board (CSB), *Anatomy of a Disaster: Explosion at BP Texas City Refinery*, published 2006.

12. Dowell, A., and Bridges, W., *More Issues with LOPA - from the Originators*, 11th GCPS, AIChE, April 2015

13. Bukowski, Julia V. and Goble, William M., Villanova University, *Analysis of Pressure Relief Valve Proof Test Data*, Process Safety Progress, AICHE, March 2009.

14. Bukowski, Julia V. and Goble, William M., Villanova University, *Analysis of Pressure Relief Valve Proof Test Data: Findings and Implications*, 10th Plant Process Safety

15. Bukowski, Julia V., *Results of Statistical Analysis of Pressure Relief Valve Proof Test Data Designed to Validate a Mechanical Parts Failure Database*, Technical Report, September, exida, Sellersville, PA, 2007.

16. Bridges, W., and Rhodes, W., *Human Factors Missing from Process Safety Management (PSM) Systems*, 17th GCPS, AIChE, April 2021

17. Bridges, W., and Rhodes, W., *Human Factors Implementation – for Plant Workers*, 17th GCPS, AIChE, April 2021

18. Bridges, W., and Tew, R., *Human Factors Elements Missing from Process Safety Management (PSM)*, 6th GCPS, AIChE, March 2010

19. Bridges, W. and Rhodes, W., *Everything you need to Know about Human Reliability and Process Safety*, 17th GCPS, AIChE, March 2019

20. Bridges, W., *LOPA and Human Reliability – Human Errors and Human IPLs*, 6th GCPS, AIChE, March 2010

21. Cutchen, S., *Safety-II — Resilience In The Face Of Abnormal Operation*, 16th GCPS, AIChE, August 2020.

22. Gertman, D.; Blackman, H.; Marble, J.; Byers, J. and Smith, C.; *The SPAR-H Human Reliability Analysis Method,* NUREG/CR-6883, U.S. Nuclear Regulatory Commission, Office of Nuclear Regulatory Research, Washington, DC, August 2005.

23. Cutchen, S., *Operational Discipline is not Follow the Procedure*, 16th GCPS, AIChE, August 2020.

Appendix A: Overview of Human Error and Human Factors

Experts typically quote that about 85% of accidents are caused by human error, though some say that except for natural disasters this figure is 100-%. However, simply attributing these incidents to "human error" without evaluating the root cause implies that the errors are inevitable, unforeseeable, and uncontrollable. Nothing could be further from the truth.

Human errors are sometimes mistakenly called procedural errors. This is not true any more than saying all equipment errors are due to design errors. People make mistakes for many reasons, but PII estimates that only about 10% of accidents due to human errors in the workplace occur because of *personal* influences, such as emotional state, health, or carelessness; most human error is due to weaknesses in the control of human factors. Over the many decades of industry research and observation in the workplace on human error, industry has come to know that human error probability depends on many factors. These factors (described in more detail in *Human Factors Missing from PSM* [18]) include those shown below (note that the percentages shown below were developed by PII after analysis of more than 15,000 process safety, safety, and operational incidents):

- Procedure accuracy and clarity (the most cited root cause of accidents):
  - A procedure typically needs to be 95% or better accuracy to help reduce human error; humans tend to compensate for the remaining 5% inaccuracies in a written procedure.
  - A procedure must clearly convey the information and the procedure should be convenient to use.
  - Checklist features – These should be used and enforced either in the procedure or in a supplemental document.
- Training, knowledge, and skills
  - Employees should be selected with the necessary skills before being hired or assigned to a department.
  - Initial Training – There must be effective training. The initial training should be demonstration-based training on each proactive task and each reactive (e.g., response to alarm) task.
  - Ongoing validation of human action is needed and usually should be repeated (in either actual performance or in drills/practice) at least once per year. For human IPLs or safeguards, the action should be demonstrated to be "fast enough" as well.
  - Documentation – the human performance should be documented and retained to demonstrate the error rates chosen are valid.
- Fitness for Duty – Includes control of many sub-factors such as fatigue (a factor in a great many accidents), stress, illness and medications, and substance abuse.
- Workload management – Too little workload and the employee becomes bored (reducing alertness, increasing distractions), while too much overwhelms the employee (increasing stress, decreasing time per task); both cases can increase human error.
- Communication – Miscommunication (of an instruction or set of instructions or of the status of a process) is the second or third most common cause of human error in the workplace. There are proven management systems for controlling communication errors (such as repeat back, use of common jargon).
- Work environment – Factors to optimize include lighting, noise, temperature, humidity, ventilation, and distractions.

- Human System Interface – Factors to control include layout of equipment, displays, controls and their integration to displays, alarm nature and control of alarm overload, labeling, color-coding, error-proofing measures, etc.
- Task complexity – Complexity of a task or job is proportional to the (1) number of choices available for making a wrong selection of similar items (such as number of similar switches, number of similar valves, number of similar size and shaped cans), (2) number of parallel tasks that may distract the worker from the task at hand (leading to either an initiating event or failure of a protection layer), (3) number of individuals involved in the task, and (4) judgment or calculation/interpolation, if required. For most chemical process environments, task complexity is typically low (one action per step), but for response actions (human IPLs) there are almost always other tasks underway when the out-of-bounds reading occurs or the alarm is activated.

In addition to the human factors listed, other considerations for use of a human as an IPL include (1) time available to perform the action and (2) physical capability to perform the action safely.

Other papers provide much more detail on each human factor and the relative weighting of each.[10, 18, 19]

These human-error causes (human factors), which in turn result from other human errors, are all directly within management's control. When using human error data for controlling initiating events (IEs) and independent protection layers (IPLs), the site should ensure that the factors above are consistently controlled over the long-term and that they are controlled to the same degree during the mode of operation that the PHA, HAZOP, What-if, FMEA, or LOPA covers. For instance, if workers are fatigued following many extra hours of work in a two-week period leading up to restart of a process, then the human error rates can increase by a factor of 10 times or 20 times during startup.[4]

Human Factor Categories and Typical Impact of Each

To minimize human error, process safety systems should address the **Human Factors Categories** (see various US NRC and US DOE standards from 1980s and 1990s)[8, 22]. Table A1 on the next page lists the key human factor categories along with multiplication factors that poor human factors can have on the base human error rates.

Table A1. SUMMARY TABLE of 10 HUMAN FACTOR CATEGORIES

Based in part on: Gertman, D.; et. al., *The SPAR-H Human Reliability Analysis Method*, NUREG/CR-6883, U.S. Nuclear Regulatory Commission, Office of Nuclear Regulatory Research, Washington, DC, August 2005[22]. PII has modified the list slightly to account for general industry data and terminology and to incorporate PII internal data.

Human Factor Category	Human Factor Issue/Level	Multiplier for Cognitive & Diagnosis Errors
Available Time (includes staffing Issues) – ***for responses only***	Inadequate time	P(failure)=100%
	Barely adequate time (≈2/3 x nominal)	10
	Nominal time (1x what is expected)	1
	Extra time (at least 2x nominal and >20 min)	0.1
	Expansive time (> 4 x nominal and > 20 min)	0.01
Stress/Stressors (includes staffing issues)	Extreme (threat stress)	5
	High (time pressures such as during a maintenance outage; issues at home, etc.)	2
	Nominal	1
Complexity & Task Design	Highly complex	5
	Moderately complex (requires more than one staff)	2
	Nominal	1
	Obvious diagnosis	0.2
Experience/Training	Low	10
	Nominal	1
	High	0.5
Procedures	Not available in the field as a reference, but should be	20
	Incomplete; missing this task or these steps	8
	Available and >90% accurate, but does not follow format rules (normal value for process industry)	3
	Good, 95% accurate, follows >90% of format rules	1
	Diagnostic/symptom oriented	1
Human-Machine Interface (includes tools)	Missing/Misleading (violates populational stereotype; including round valve handle is facing away from worker)	20
	Poor or hard to find the right device; in the head calc	10
	Some unclear labels or displays	2
	Good	1
Fitness for Duty	Unfit (high fatigue level (>80 hr/wk or >20 hr/day, no day off in 7-day period; or illness, etc.)	20
	Highly degraded fitness (high fatigue such as >15 hr/day, illness, injury, etc.)	10
	Degraded Fitness (>12 hr day and >72 hr/wk)	5
	Slight fatigue (>8 hr per day; ***normal value for process*** industry)	2
	Nominal	1
Work Processes & Supervision	Poor	2
	Nominal	1
	Good	0.8
Work Environment	Extreme	5
	Good	1
Communication	No communication or system interference/damage	10
	No standard for verbal communication rules ***(normal*** value for process industry)	3
	Well implemented and practiced standard	1

If all of these human error factors are controlled very well, then the optimized human error rates listed in Section 4 of this document are achievable. Alternatively, if one or more of these factors are compromised, the human error rate will increase by the values shown in this table. As an example, an individual whose fitness for duty rating is unfit due to the excessive work hours shown in the table will make errors at a rate 20 times greater than an individual whose fitness for duty is normal.

With excellent control of each of the human factors listed above, a company can begin to approach the lower limits that have been observed for human error. These lower limits are about:

- **1 mistake in 100 steps for most procedures-based tasks** (such as starting up a process unit), a little less for a routine (daily) task that becomes almost a reflex
- **1 in 10 chance or a little better for diagnosis and response to a critical alarm**

Excellent control requires superior design and implementation of management systems, which is enabled through a thorough understanding of these factors.

619469 2021 AIChE Spring Meeting & 17th GCPS Thursday Keynote: Industry Perspectives on Learning and Innovation

Abstract:

AIChE has launched the Institute for Learning and Innovation; the vision is to be the horizontal connector between universities and industry for continual excellence in learning for students and working professionals. This panel features industry leaders who will share their observations on industry needs for workforce of the future. We plan to have this be an interactive discussion with those attending.

SPRING21 +17TH GCPS
A Joint AIChE and CCPS Meeting

Measuring Vital Signs of Process Safety Culture

Hamza Hameed
Senior Process Safety Engineer
Fatima Fertilizer Company Limited, Pakistan
hamza.hameed@fatima-group.com

Muhammad Ahsan Sarfraz
Manager Health, Safety and Environment
Fatima Fertilizer Company Limited, Pakistan
Ahsan.sarfraz@fatima-group.com

Prepared for Presentation at
American Institute of Chemical Engineers
2020 Spring Meeting and 17th Global Congress on Process Safety
Virtual
April 18 - 22, 2021

Measuring Vital Signs of Process Safety Culture

Hamza Hameed
Senior Process Safety Engineer
Fatima Fertilizer Company Limited, Pakistan
hamza.hameed@fatima-group.com

Muhammad Ahsan Sarfraz
Manager Health, Safety and Environment
Fatima Fertilizer Company Limited, Pakistan
Ahsan.sarfraz@fatima-group.com

Keywords: Process Safety Culture & its core principles, Sense of Vulnerability, Normalization of Deviance, Fostering Mutual Trust, Assessing & advancing Process Safety Culture, Deferring to Expertise, Establishing an Imperative for Safety,

Abstract

In line with Fatima Group's vision and HSE policy, Fatima Fertilizers Company Limited (FFCL) stipulated a Process Safety Management program (PSM) as the major corner-stone. It was purely a self-initiative to implement this model rather following any regulatory requirement. At Fatima Fertilizer Company Limited, Process Safety Performance is tracked using Global Method of Leading and Lagging Indicators. Leading Indicators are a set of data that measures performance across various elements of Process Safety. In 2020, in order to uplift Process Safety performance at FFCL, we invested in activities keeping in view Core Principles of Process Safety Culture. While analyzing Leading Indicators for 2019, FFCL performance across various elements was linked with Core principles of Process Safety; this helped to identify strong and weak areas of our culture. To supplement this analysis in pursuit of strong Process Safety Culture, another study was conducted with title **"What is it in for me, Two Decades of Process Safety Beacons".** All PS beacons till date were critically reviewed and relevant information was extracted. Main objective of this activity was to learn from past mistakes and strengthen Process Safety elements

1 Introduction

Fatima Fertilizer Company Limited (FFCL) is the first and the only green field project which has materialized under the 2001 Fertilizer Policy of the Government of Pakistan. Fatima Fertilizer Limited has successfully implemented ISO 45001:2018, ISO 9001:2015 and ISO 14001:2015 as Integrated Management System (IMS). Moreover, Fatima Fertilizer Limited has implemented DuPont 22 element model of Process Safety Management (PSM). This journey started in 2012 and in span of 5 years, FFCL site achieved a rating of 04 equivalents to Excellence (Maximum rating 05 equivalent to World Class). FFCL site is also IFA Protect & Sustain certified site. Our continuous focus on Health and Process Safety is evident from our Safe Million Man-Hours figure (current number is 59 Million).Last LTI at FFCL was recorded in 2012. The Total Recordable Injury Rate (TRIR) remained zero in 2019 and 0.03 in 2020. FFCL has also achieved the Royal Society for the Prevention of Accidents (RoSPA-UK) health and safety award (Gold) straight for last 04 years. FFCL Site has also the honor of achieving 04 British Safety International Safety Awards in a row since 2014 and achieved distinction in 2019. FFCL laboratory is accredited with ISO/IEC-17025 which is 1st fertilizer lab of Pakistan. Furthermore, FFCL Learning and Development center also known as training center is also certified with ISO-29990 which makes FFCL 1st manufacturing site in the region to have this honor. [1]

This paper will describe FFCL efforts to uplift Process Safety Culture at site. FFCL Process Safety Management journey started in 2012 after a fire incident at an Ammonia plant. Rigorous efforts were made to reduce Process Safety Incidents at the site. Acknowledging the fact that cultural aspect of Process Safety affects all other elements of Process Safety and its improvement is one of the most challenging and time intensive activities, FFCL invested in activities to uplift Process Safety Culture. Moreover, all these activities were aligned with Centre for Chemical Process Safety (CCPS) Process Safety Culture Core Principles.

1.1. Facility Description

FFCL is one of Pakistan's largest fertilizer manufacturing complexes that process natural resources like natural gas, rock phosphate, limestone and water into vital products for farmers and industrial customers. FFCL final products include urea, calcium ammonium nitrate (CAN) and nitro phosphate (NP), while Ammonia and Nitric Acid are the intermediate products. The complex contains integrated utilities section for all 5 production units (Calcium ammonium nitrate, Urea, Nitro phosphate, Nitric Acid and Ammonia) and 1 bagging unit for final product handling and dispatch. The fertilizer complex is a fully integrated production facility occupying 947 acres of land and a total of 1886 employee's including 215 management, 556 staff and 1115 contract employees.

The parent unit of the complex is a 1700 MTPD Ammonia plant which is currently operational using natural gas as the basic raw material. This ammonia plant was relocated from Rozenburg, Netherlands to Pakistan in year 2007. Prior to relocation of the ammonia plant it had gone through several revamps and energy conversation projects resulting in an increase in production capacity from 1350 MTPD to 1650 MTPD. The plant was shut down in 2000 and purchased by Fatima group in year 2006 and relocated to Pakistan in year 2007.

The ammonia and its by-product carbon dioxide are used for the production of 1500 MTPD prilled urea. The urea plant is a Stamicarbon Urea 2000+ technology new plant and was commissioned in 2010. The nitric acid production facility has a capacity of 1500 MTPD (based on 100% conc.), which along with ammonia and limestone, is utilized for the production of 1200 MTPD prilled nitro phosphate and 1400 MTPD granulated calcium ammonium nitrate. The NP plant is designed by Chemical Fertilizer Industry and it was commissioned in June 2011. The Ammonia, CAN and NA plants are relocated plants. The complex is self- sufficient in power generation and also supplies a residential colony. Figure-1 below illustrates the main process units at the FFCL plant. [1]

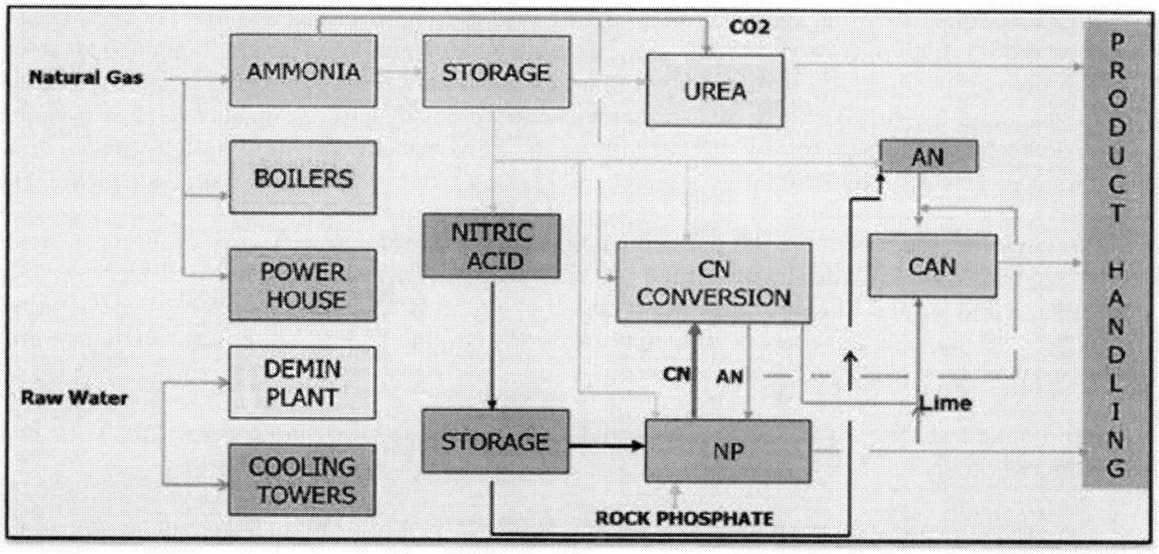

Figure 1. FFCL Flow Diagram- Raw Materials Conversion into Products

1.2. PSM Implementation and Growth History at FFCL Site

FFCL established in 2007 and conventional Safety Management systems were adopted initially which mainly worked around personnel safety performances while process safety was not the prime focus area. No formal process safety management system was launched, and the organization was still young without a rich culture and legacy. Employees from different backgrounds and cultures had joined the organization so overall a very lean safety culture existed. There were no legislative requirements in the region for process safety systems rather legislation are largely focused on labor laws. Consequently the overall site safety performance remained unsatisfactory with major setbacks in the form of severe process safety incidents (major fires & Tier Level-1 Process Safety incidents). Below are a few glimpses.

- FFCL had two fatalities in 2010 and one in 2011.One fatality occurred due to electrocution as there were no regime of electric tools testing and unskilled person tried to rectify the faulty electric asset. The second fatality occurred due to slippage of person in created opening.
- 22 recordable injuries were reported in 2011. Overall number was 70 including both 2010 and 2011. 70% of these injuries were related to contractors.
- 4 Lost Time Injuries, 1 Restricted Work Case, 2 Medical Treatment Cases & 48 First Aid Cases were reported in 2012.
- High number of fires with 29 process and non-process fires occurred in 2012
- Elevated Total Recordable Injury Rate of 0.51 (per 200,000 hours).

Major process safety incidents like an ammonia syngas compressor fire and operational upsets like Catacarb incident occurred due to which the plant remained unavailable for more than 40 days. Following the 2012 series of process safety upsets on the ammonia plant, site management analyzed the situations and identified inadequate process safety management systems as the root cause. Subsequently it was agreed to align exiting systems and practices with Best in class Process safety model (DuPont PSM model) for sustainable business model and to enhance safety, reliability and productivity of site. The whole PSM system was implemented at site in less than four years and the first 3rd party assessment was carried out by DuPont in May 2015 which rated FFCL at 3.5 from scale 1 to 5 which meant that the PSM system was well implemented. The 2nd assessment was done by DuPont in November 2017 and rated FFCL site at Excellence level. [1]

Following the successful implementation and external Third Party Audits of Process Safety Management at FFCL site, focus shifted towards program sustainability. While External Audits are planned at a span of every three years, internal audits (1st Party & 2nd Party) were regularly conducted at site. All these audits pointed towards an alarming fact of **A Stagnant Process Safety Culture.**

After detailed analysis by FFCL Process Safety Engineers, the following causes were at the heart of Process Safety Culture Stagnation:

- Since worker injury rate was at decline at FFCL, workers and management assumed that this was also indicating that the risk of Low Frequency, High Consequence Process Safety was at decline.
- As injuries are easily classified and injury rates calculated through Total Recordable Injury Rate (TRIR), focus sometimes inadvertently shifts towards Personal Safety instead of Process Safety.
- Process Safety Management was implemented as One-Time project. Efforts were not put to embed in FFCL working culture
- FFCL strong Personal Safety record had led its management to complacency. The absence of a major incident had lessened FFCL Sense of Vulnerability. FFCL Statistics had continued to demonstrate that our safety system were better than others working around us.
- FFCL Process Safety Engineers had failed to demonstrate the significance of Process Safety Culture. Although we had developed comprehensive PSM Standard Operating Procedures, training plans and audit regimes, our discussion with FFCL management and employees did not include cultural aspects related to all PSM Elements

2 Focusing Process Safety Culture

In early part of 2019, this grey area of Process Safety was ultimately realized and in second half of 2019, efforts were initiated to uplift Process Safety culture. To ensure we moved towards the right direction, we decided to consult Global Practices. After consulting multiple practices and literature, we narrowed our approach to CCPS Process Safety Culture Core Principles. In the next step, activities aligned with these core principles were agreed upon and were later executed.

2.1 Learn to Assess and Advance Process Safety Culture: [3]

2.1.1 Leading and Lagging Indicators Analysis:

At FFCL, Process Safety Performance is tracked through a set of Leading and Lagging Indicators. This set is summarized and presented to site leadership team in monthly Executive Health, Safety and Environment Committee. FFCL Leading Indicators track site performance across different PSM elements and include followings

- Process Safety Action Items Compliance
    - Incident Investigation Reports and Recommendations' Compliance
    - Quarterly Emergency Drills' Recommendations Compliance
    - Process Hazard Analysis (PHA) and Recommendations' Compliance
    - Process Safety Management Audit Recommendations' Compliance
    - Permit to Work Quality Audit Score
- Mechanical Integrity Leading Indicators
- Safety Instrumented System Integrity
- Management of Change Audits' Score
- Operating and Maintenance Procedure Revision Compliance
- Management Safety Audit and Recommendations' Compliance
- Executive Health, Safety and Environment Committee Recommendations' Compliance
- Un-Safe and Un-Safe conditions' Reporting
- Job Cycle Check Compliance

While lagging Indicator matrix, defined a debit score for followings

- Process and Non-Process Fires
- Process Safety Incidents
- Repeat Incidents
- Environment Incidents
- All categories of Injuries

At start of each year, an annual analysis was conducted by FFCL Process Safety Engineers to highlight strong and weak areas of Process Safety Management and based on this analysis Leading and Lagging Indicators were revised. In 2019, while conducting Leading and Lagging Indicators' Analysis, strong and weak aspects of Process Safety Culture were examined. Figure 2 below summarizes weak areas of Process Safety associated with the element of "Process Hazard Analysis". FFCL, PHA Standard Operating Procedure, categorizes 4 levels of PHA Recommendations based upon associated risk. Level-1 PHA recommendations address highest risk of the concerned area while Level-IV PHA recommendations are associated with the lowest risks. Level 1 & 2 PHA recommendations have different hierarchy of approval and waiver process (approved by highest authority of any department) than Level 3 & 4 (approved by line managers)

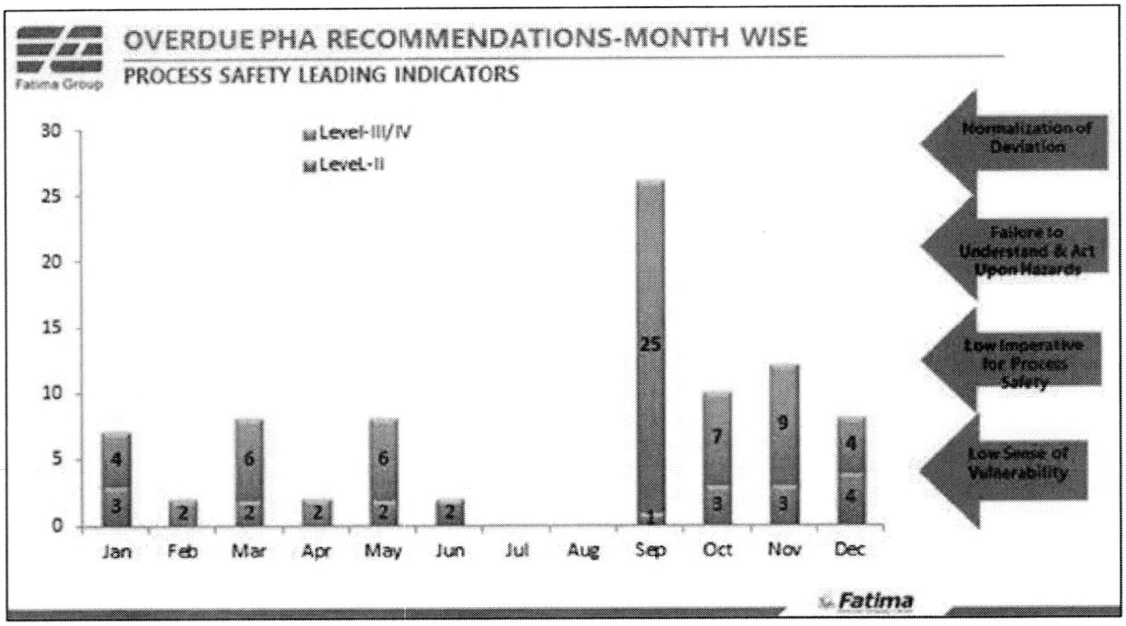

Figure 2 Cultural Aspects Related With Process Hazard Analysis Recommendations

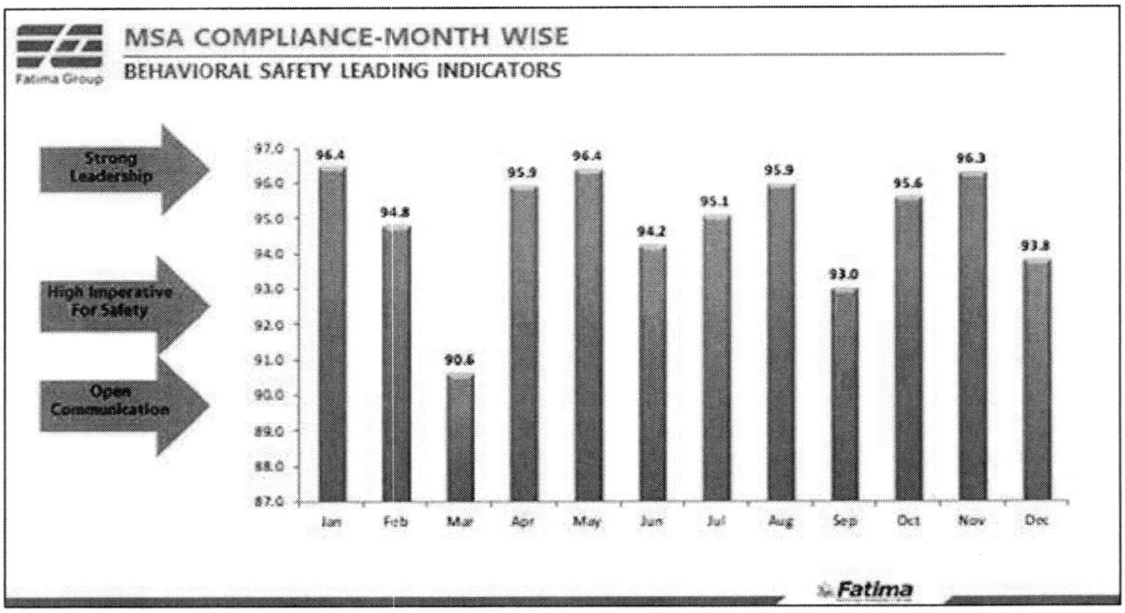

Figure 3 Cultural Aspects Related With Management Safety Audits (MSA) Compliance

This method of Leading and Lagging Analysis unlocked an insight into cultural aspects of Process Safety Activities. Action Plans generated based on this analysis (Some examples are mentioned below) ensured weaker cultural areas were focused on in the coming years.

- Risk Based (High Risk Areas) Management Safety Audit Schedule for Managers, Increased Weightage of Executive Health, Safety and Environment Committee Recommendation ⟶ **Providing Strong Leadership**
- Risk Based Scoring in Leading Indicators for Process Hazard Analysis Recommendations, Increased Score Debit in Leading Indicator for each Temporary Repair ⟶ **Normalization of Deviance**

2.2 Fostering Mutual Trust/ Ensuring Open and Frank Communication: [3]

Keeping in mind the famous saying "What gets measured, Gets Managed", a comprehensive Process Safety Survey was compiled. Sets of close ended questions from all core principles were conceived and it was completed by all FFCL employees either through email or in hard form. Second phase of the survey included evaluating Survey Results and focusing on major concerns, results' communication and sharing frequent feedback with the employees. This whole scheme was targeted to receive true feedback on FFCL Process Safety Culture and to gain the trust of employees. All questions in the survey were divided into 9 core principles of Process Safety. Specimens from each core principle is enlisted in below table

Table 1: Process Safety Culture Survey Specimen

Core Principle	Question	Yes/No
Establish an Imperative for Safety	Is the PSMS a detailed set of management system procedures that represents a "paper only" program that sits on the shelf, or has it actually been implemented and is it being used?	
Providing Strong Leadership	Is process safety an agenda item or is it considered at high level meetings on a regular basis (not just after an incident)?	
Foster Mutual Trust	Does the sharing of information that will reduce safety risks occur without fear of punishment?	
Ensure Open and Frank Communications	Are those bearing negative safety-related news required to "prove it is unsafe?"	
Maintain a Sense of Vulnerability	Do risk analyses routinely eliminate proposed safeguards under the banner of "double jeopardy?"	
Understand and Act Upon Hazards/Risks	Are the appropriate resources applied to the hazard/risk assessment process?	

Empower Individuals to Successfully Fulfill their Safety	Are employees empowered to suggest or initiate improvements?	
Combating the Normalization of Deviance	Is there undue peer pressure or organizational pressure to work extensive amounts of overtime?	
Learn to Assess and Advance the Culture	Has the organization established and nurtured a questioning and learning environment?	

2.3 Providing Strong Leadership [3]

The leadership team's role in strengthening culture is of utmost importance. Leadership team's communication must lay down significance of Process Safety. Acknowledging this fact, a half day workshop on Process Safety was arranged for FFCL Leaders. The initial idea was to include the top layer of FFCL managers but bearing in mind the role of line managers, participation was widened. This workshop was led by Fatima Group Corporate HSE Manager with more than 50 participants. Key topics of workshop included

- Process Safety Leadership Basics
- Characteristics of Leadership and Management in Process Safety Culture
- Consistence of Process Safety Messages
- Case Studies of Past Incidents
- Turnover of Leadership, Succession Planning and Organizational Management of Change

By dedicating the site leadership team's time for a discussion on Process Safety a strong message was sent across the site that "Process Safety will always take the front seat"

2.4 Maintaining a Sense of Vulnerability [3]

A high sense of Vulnerability is an indication of Strong Process Safety Culture. With strong safety statistics and lack of major incident, a sense of complacency was being felt in working patterns and communications. To overcome this problem, a campaign with title **"On This Day in History"** was run at FFCL site. Details of major process safety incidents around the world, FFCL site and other Fatima Group sites were collected and on the day of any particular incident, a memorial flyer was shared with all Fatima Group Sites. These flyers were discussed in Weekly Safety Talks. Total of 58 incidents were shared on their anniversaries with either of below messages:

- IF IT HAPPENED THERE, CAN IT HAPPEN HERE?
- IF IT HAPPENED THEN, CAN IT HAPPEN AGAIN?

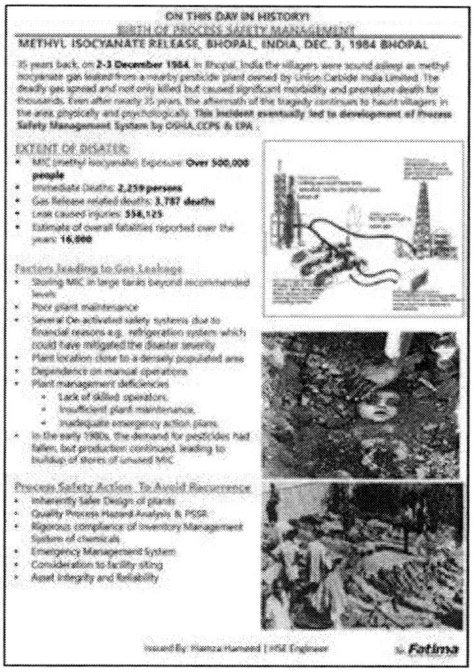

Figure 4 & 5: On This Day in History Flyer on Bhopal Tragedy and FFCL Incident

All "On This Day in History" flyers were discussed in all weekly Safety Meetings (D-Level Talk). This valuable discussion on past incidents provided an excellent opportunity for learning and as well as uplifted Sense of Vulnerability across site. Considering the benefits of this exercise, this practice was later initiated in other Fertilizer Sites of Fatima Group in 2021.

2.5 Defer to Expertise [3]

To ensure all relevant process safety and technical information are communicated to relevant individuals, a detailed study of Process Safety Beacons was conducted in 2020. The assigned individual studied 2 decades (total 226 beacons) of CCPS process safety beacons and developed an action plan for each beacon. The main idea was to learn from mistakes of others and to share Process Safety Beacons with concerned interfaces only. Learning from beacons was divided into 4 categories

- For Discussion in weekly safety talks
- Conducting Field Audits
- Conducting Mock Drill Exercises
- Conducting detailed Trainings

Each beacon covers a Process Safety or technical topic pertinent to a specific interface. For example, Feb-2012 Beacon (Mechanical Integrity-The Bolts Are Too Short) is related to Maintenance teams or Sept-2008 (Plant Security) is related to Security Personnel. Through this study, a matrix was developed that segregated information sharing according to their end users.

Below table includes the number of Process Safety Beacons against each department discussed in their weekly safety talks.

Table 2: Number of Process Safety Beacon Distribution w.r.t End Users

Topics	Number of Beacons to be Shared in Weekly Safety Talks
Relevant to all site	33
Production Related	162
Maintenance Related	31
Electrical and Instrument Related	15
Laboratory Related	06
Plant Security Related	04
Emergency Response Team & Incident Commander Related	11
Material Warehouse Related	21
Technical Services Related	09

2.6 Empowering Individuals to Successfully Fulfill their Process Safety Responsibilities [3]

To ensure all Process safety activities are performed as intended, competency of individuals is one of the key factors. After successful implementation of Process Safety Management, a Training Need Analysis of all employees was conducted w.r.t Process Safety. Training programs for most commonly used Process Safety Activities were developed and external trainers were hired. For instance, Root Cause Analysis and Process Hazard Analysis trainings were delivered by DuPont Specialist trainers. Empowering individuals to fulfill their Process Safety tasks demand high level of expertise. This has been achieved by conducting Process Safety Trainings on cyclic basis. For example FFCL Annual Training Calendar contains below trainings each year

- Process Hazard Analysis- Two day Training
- Incident Investigation and Root Cause Analysis- Two Day Training
- Process Safety Management Workshop- Two Day Training
- Pre-Startup Safety Review- Half Day Training
- FFCL Permit to Work System- One Day Training
- Emergency Response Practical and Theoretical Trainings

Through this cycle of training and refresher sessions, employee's and contractor's process safety knowledge base is strengthened and hence is reflected in strong process safety performances. For example Root Cause Analysis training contains numerous individual and group exercises for finding correct Root Cause, Using Why-Tree technique for Root Cause and generating correct Recommendations at the end. For this particular training, average score for Incident Review increased from **80 to 85%.** This review is conducted by a committee called "Incident Report Review Committee" and is led by a senior manager. This committee randomly selects Detail Investigation Reports and evaluates on a fixed criteria. Some trainings' also include Pre and Post Training tests to quantify improvement in knowledge base of participants.

2.7 Establishing an Imperative for Safety [3]

For a production site, difference of opinion between production and safety groups is a never-ending debate. Prioritizing safety over production depicts a strong safety culture. FFCL is no exception w.r.t this discussion. During initial phases of PSM implementation, numerous events were reported when a process safety activity was by-passed due to production pressures e.g. deliberately skipping a Pre-Start-up Safety Review or delaying a Process Hazard Analysis study etc. All such cultural issues have been gradually addressed by carefully changing some practices that indicated a High Imperative of Process Safety. Few examples include

- A norm was introduced to start all communication meetings for site employees with a SAFETY PAUSE, which can be a short video, a short safety message or a recent safety incident. In 2020, all meetings initiated with a safety pause on Pandemic.
- In 2020, sequence of Plant Morning Meetings was changed with HSE manager starting the session. In these meetings, all on-going major process safety concerns or recent incident are shared with site, giving a strong message across site that Our Work Starts with Safety.
- Similarly Management Review Sessions with CEO are started with Safety Stats that include Current Safe Million Man-Hours Number, Total Recordable Incident Rate and Summary of running year incident.
- Like any other organization, at FFCL incentives are given for achieving Production Targets. Applying same philosophy to safety matters, HSE rewards are distributed for achieving safety related milestones for example gifts distribution across site for achieving each safe million man-hour. Furthermore, in 2019 new categories were added in FFCL Achievers Event for an example Best investigation for an incident conducted by a team.

All these steps have helped to create an aura of Process Safety at FFCL.

2.8 Combatting the Normalization of Deviance [3]

From the start of 2020, FFCL HSE section has put special efforts to highlight areas where site performances have been continuously below par. At FFCL, Process safety performance is monitored through Leading and Lagging Indicators, presented to site top management each month. Process safety engineers continuously look out for abnormal trends across PSM elements, and alarm is raised in monthly Executive HSE Committee to site leaders. By timely raising concerns, normalization of deviances is avoided. For an example in 2020, serious observations relating to Work at Height were being repeatedly reported. A detailed Permit to Work analysis was presented to top management in September, which ultimately led to special focus towards Scaffolding practices. An action plan was developed that included Scaffolding Procedure revision and training of all scaffolding inspectors, supervisors and workers. Below figures are other two examples of concerns being raised in Executive HSE Committee Reviews

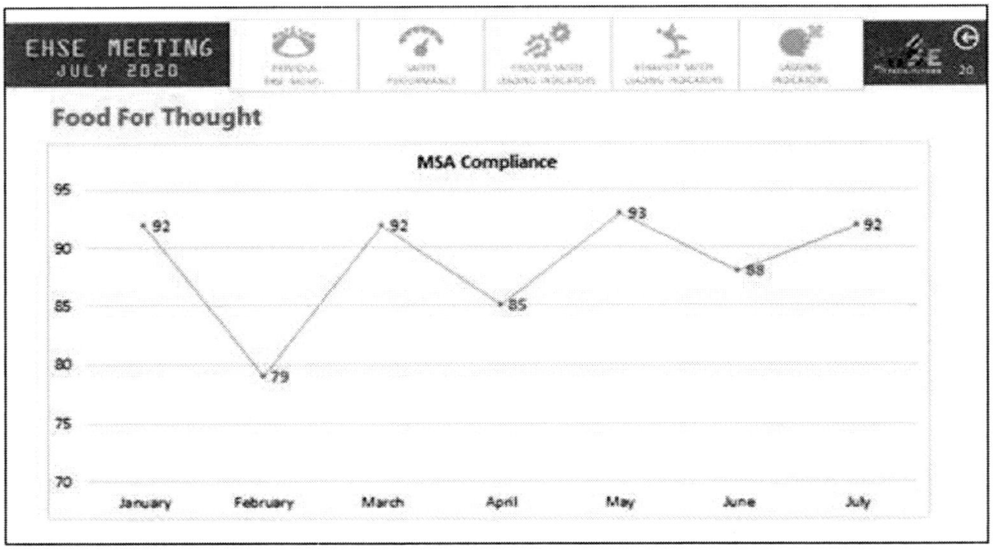

Figure 6: Management Safety Audits Varying Compliance Trend (July-2020)

Figure 6 is extracted from Executive Health, Safety and Environment Committee Review (July-2020). Through this data, a concern was brought in notice of senior management that Compliance of Management Safety Audits is not consistently good. Management Safety Audits are performed by all engineers once per month and historically has been one of key strengths of our Process Safety System. By raising timely concern, a message was communicated across all site (through their managers) that this area needs immediate focus. Improvement was observed immediately in proceeding months

3 Conclusion

Investing in Process Safety Culture has thus far helped to uplift Process Safety Performance at FFCL site. One lesson that FFCL has learnt over years about Process Safety is that Cultural issues are the most difficult tasks to deal with. Over the years different definitions have emerged for Process Safety Culture, most commonly used are “what people do when their boss is not around” or “The combination of group values and behaviors that determine The Way Process Safety is managed”

Cultural issues can go un-noticed and may suddenly result in a significant loss. One of the reasons for these cultural issues going un-noticed is difficulty in measuring an organizations’ cultural status. It took FFCL several years to determine cultural status of site by agreeing upon vital signs of Process Safety Culture. For example Leading Indicators’ Analysis unlocked strengths and weakness associated with process safety activities for example Process Hazard Analysis,

Management Safety Audits, Pre-Start up Safety reviews etc. Other activities like Process Safety Survey gave qualitative analysis of Process Safety Culture. Some other cultural aspects are harder to predict and afterwards rectify for example Complacency or Lower Sense of Vulnerability. Process safety incidents over the years have taught one important lesson is there is no such thing as "Stagnant Culture" either culture has an upward trend or a downward. To carter this cultural aspect, "On This Day in History" has helped FFCL to reduce complacency at all levels of organization. Since, there are so many variables associated with Process Safety culture, each organization will face different challenges and accordingly they need to work out Vital Signs of Process Safety Culture.
Below are some valuable lessons learnt regarding Process Safety after implementing above mentioned activities

- "Out of the box thinking" in Process Safety is equally important like any other aspect of life.
- A stagnant Process Safety Culture is the one on decline. Employees must be continuously engaged in Process Safety activities to get the best results.
- A strong Process Safety Culture sees off difficult times unharmed for example unplanned extended shutdowns, mergers, acquisitions and even economic crisis

This cultural shift has started to show its positive result and improved results are now shown at various fronts.

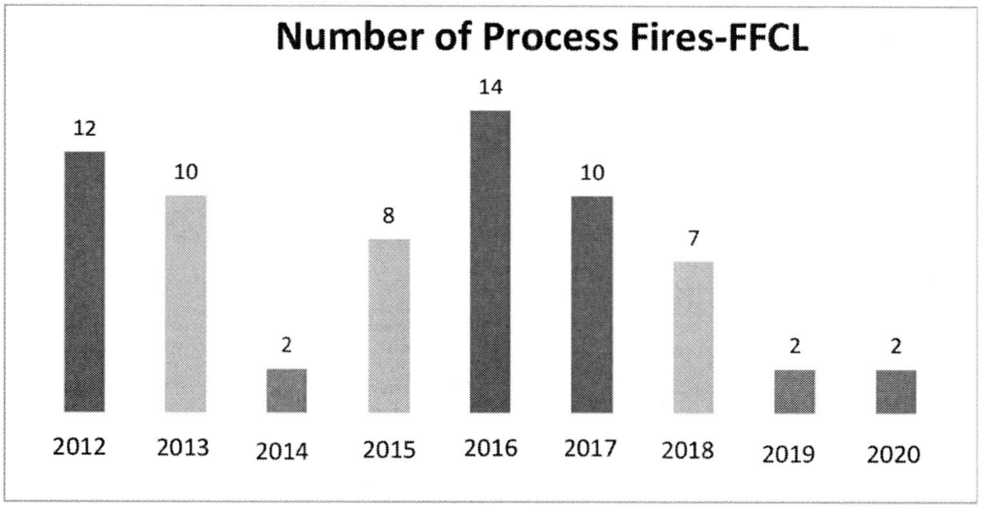

Figure 7: Number of Process Fires- FFCL

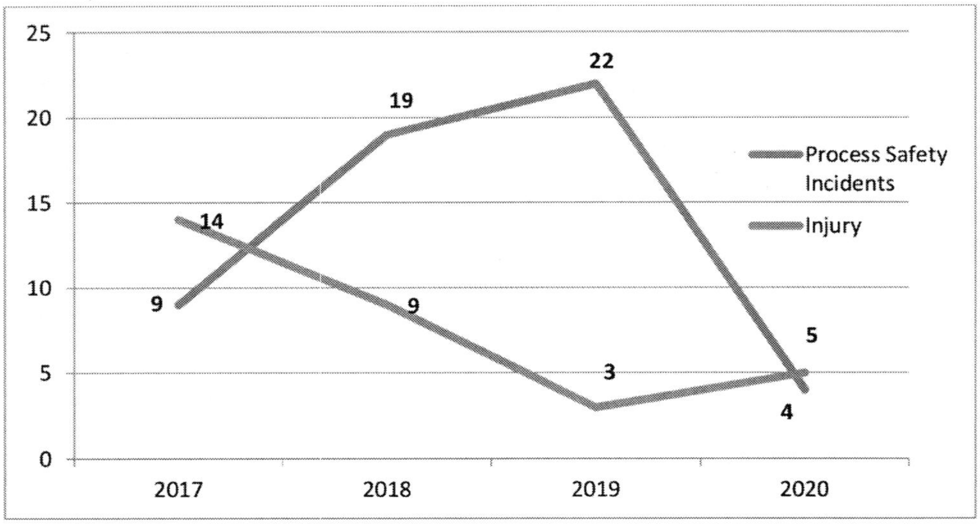

Figure 8: Number of Process Safety Incidents and Injuries- FFCL

4 References

[1] Ahsan Sarfaraz/ Usman Ali Mughal, "50 Million Safe Man Hours without a Zero Loss Time Injury," in 16TH Global Conference on Process Safety, 2020.

[2] Guidelines for Risk Based Process Safety, First Edition. CCPS. ©2007 AIChE. Published by John Wiley & Sons, Inc.

[3] Essential Practices for Creating, Strengthening, and Sustaining Process Safety Culture, First Edition. CCPS. ©2018 AIChE. Published 2018 by John Wiley & Sons, Inc.

In Process Safety - Is 14 audit findings a good vital sign

Larry Lepard, CPSA
Principal Consultant - ERM
9825 Kenwood Road, Suite 100, Cincinnati, OH 45242
Larry.lepard@erm.com

Prepared for Presentation at
American Institute of Chemical Engineers
2020 Spring Meeting and 17th Global Congress on Process Safety
Virtual
April 18 - 22, 2021

In Process Safety - Is 14 audit findings a good vital sign

Larry Lepard, CPSA
Principal Consultant - ERM
9825 Kenwood Road, Suite 100, Cincinnati, OH 45242
Larry.lepard@erm.com

Keywords: Process Safety, Compliance Audit, Metrics, Vital Signs

Abstract

PSM compliance audits are required every three years. What do the audit results tell you about your process safety program? Is the number of audit findings a useful vital sign?

It is critical to know how a company performs, views and uses the results of a process safety compliance audit. This can greatly affect their overall risk management and therefore the protection of their employees and neighbors. There can be a tendency to look at the number of audit findings and not the potential severity.

Some companies focus strictly on the numbers, try to compare results between different operations or incorporate the results into employee performance reviews and/or bonus payouts. These practices may have unintended negative consequences.

The author will share lessons learned, that were made while conducting hundreds of compliance audits, and will discuss how leading organizations manage their audit findings and use the results to improve performance and reduce risks.

1 Introduction

Under the Process Safety Management (PSM) Regulation, 29 CFR 1910.119, every covered facility is required to certify that they have evaluated compliance. This is accomplished by conducting a compliance audit of the provisions of the PSM regulations at least every three years. The main purpose of this exercise is to verify that the procedures and practices developed under the PSM standard are adequate and are being followed.

An audit is a fundamental part of an effective PSM program because its purpose is to verify that "systems to manage process safety" are in place and functioning effectively, and to take corrective action when findings indicate that is warranted [1]. The goal of a PSM program is to prevent process safety related incidents from occurring; or at the least, to minimize the potential frequency and severity of incidents should they occur. Process safety incidents can result in a toxic release, fire or explosion. These incidents can cause deaths, severe injuries, environmental impacts and lost production. Just as process safety incidents are rarely caused by a single event, rather a combination or a series of multiple events and failures, all the elements of a PSM program are designed to work together to prevent incidents. As such, an effective audit must evaluate all the elements to identify where there are gaps. Gaps in this sense can refer to regulatory non-compliance or not implementing what are generally accepted best industry practices. The conclusion of a compliance audit is generally an audit report that identifies findings (gaps) and recommendations for improvement. What a company does with the report and how they manage the recommendations can greatly affect the overall success of the PSM program.

2 PSM Program and Compliance Audits

The OSHA PSM regulation was issued in 1992 consequential to which, the PSM compliance audits have been required since 1995. Unlike some of OSHA's standards, which prescribe precisely what employers must do to comply, PSM is a "performance-based" standard - it outlines key features of existing safety programs for controlling highly hazardous chemicals, and employers have the flexibility to tailor their safety programs to the unique conditions at their facilities.
Specifically, the required process safety elements as per the OSHA PSM standard (29 CFR 1910.119), are Employee Participation, Process Safety Information (PSI), Process Hazard Analysis (PHA), Operating Procedures (OP or SOP), Training, Contractors, Pre-Startup Safety Review (PSSR), Mechanical Integrity (MI), Hot Work Permit (HW), Management of Change (MOC), Incident Investigation, Compliance Audits and Trade Secrets [2]. There are many guidance documents available describing how to develop and implement an effective PSM program. The following are examples: OSHA 3908-03 2017 Process Safety Management for Small Businesses, OSHA 3132 Process Safety Management, CCPS Guidelines for Implementing Process Safety Management, and the IIAR Process Safety Management Guidelines for Ammonia Refrigeration. In general, sites are expected to develop policies and procedures that describe what will be done to comply with each element and provide training to all affected employees and contractors as necessary for implementation. Based on the author's experience and discussion with other PSM professionals, the golden rule for PSM programs is "Say what you do and do what you say". A key piece to a successful audit is having documentation to demonstrate that you have done what you say.

This paper does not focus on PSM program development but rather on the compliance audit element and the role this plays in the overall health of the PSM program. The compliance audit is intended to evaluate both the design and implementation of the PSM program on a periodic basis [3]. It is a measure of whether the program is properly developed, implemented and functioning as intended. A good audit program will identify areas of non-compliance, opportunities for improvement and should help to prioritize corrective actions. All companies have limited resources so there is a need to apply those resources where they will make the biggest impact. A successful audit process will continually improve the overall PSM program. Successful PSM programs have been shown to reduce injuries, regulatory non-compliance, production downtime and the associated costs.

3 What is required in a PSM compliance audit

The regulatory requirements for this element are relatively brief. The regulatory language from 29 CFR 1910.119(o) [2] related to compliance audits is as follows:

1. Employers shall certify that they have evaluated compliance with the provisions of this section at least every three years to verify that the procedures and practices developed under the standard are adequate and are being followed.
2. The compliance audit shall be conducted by at least one person knowledgeable in the process.
3. A report of the findings of the audit shall be developed.
4. The employer shall promptly determine and document an appropriate response to each of the findings of the compliance audit, and document that deficiencies have been corrected.
5. Employers shall retain the two (2) most recent compliance audit reports.

Again, there are many guidance documents available describing how to develop and implement an audit program. The CCPS Guidelines for Auditing Process Safety Management Systems [1] alone is a 900-page document. As such, this paper guides the readers to a successful audit through identification of some key points and sharing of lessons learned. It does not tell the reader everything they need to know to develop and implement an audit program or how to be a successful auditor.

The following are minimal requirements and key points related to each of the required items from above.

Requirement 1 – As the standard requires the employer to "certify" they have evaluated compliance, there needs to be a certification statement signed by the person responsible for PSM implementation at the facility. It is common practice to include a certification statement in the audit report or to attach a certification letter to the report. Audits and certifications must occur at least every three years. A facility may choose to conduct audits more frequently than this. Some companies choose to develop audit frequency requirements based on factors such as the number of repeat findings, high-risk findings &/or other factors.

Requirement 2 – The audit can be conducted by a single person provided that person is knowledgeable in the process. This means the person must have knowledge / experience with the covered process(es) at the facility and knowledge in conducting PSM audits. The selection of effective audit team members is critical to the success of the program. Team members should be chosen for their experience, knowledge, and training and should be familiar with the processes and with auditing techniques, practices and procedures. The size of the team will vary depending on the size and complexity of the process under consideration. The audit should be conducted or led by a person knowledgeable in audit techniques and who is impartial towards the facility or area being audited [3]. It is typical to have members with varied experience on the team, some members having more process knowledge and other members being stronger in regulatory requirements and auditing process. The audit team can consist entirely of personnel from the site or company, third party auditors or a combination of both. In my experience, a combination of internal and third party audit members can produce the best results. Internal auditors generally have a better understanding of the covered process and company requirements where third party auditors may have a better (at least different) understanding of legal requirements, can bring experience / examples from other companies and industries and can offer a fresh set of eyes.

Requirement 3 – Some form of a formal report that identifies the findings must be developed at the conclusion of the audit. An audit report can be as simple as a findings table or can be quite complex. Some companies require the completion of an audit protocol, with a set of questions for each element, which must be answered and included as part of the report. In general, audit reports typically include some form of narrative about the facility, the audit method, audit participants, description of risk ranking (if used), a summary or conclusion of the results and a detailed findings table. It is common practice for the findings table to include the regulatory or company citation, a description of the requirement, a description of the issue / gap observed and recommended corrective actions.

Requirement 4 – A corrective action plan must be developed to document what will be done to address each of the findings and document that deficiencies have been corrected. While it is not specifically stated or required in this section, it is best practice to follow the same requirements as that for PHA recommendations. As such, the corrective action plan should identify the issue, what will be done, the expected completion date and the actual completion date. It is very important to maintain documentation necessary to demonstrate that the stated corrective action was completed. Per Appendix C to §1910.119 -- Compliance Guidelines and Recommendations for Process Safety Management (Non-mandatory) "Corrective action is one of the most important parts of the audit. It includes not only addressing the identified deficiencies, but also planning, follow-up, and documentation. The corrective action process normally begins with a management review of the audit findings. The purpose of this review is to determine what actions are appropriate, and to establish priorities, timetables, resource allocations and requirements and responsibilities. In some cases, corrective action may involve a simple change in procedure or minor maintenance effort to remedy the concern. Management of change procedures need to be used, as appropriate, even for what may seem to be a minor change. Many of the deficiencies can be acted on promptly, while some may require engineering studies or in-depth review of actual procedures and practices. There may be instances where no action is necessary and this is a valid response to an audit finding. All actions taken, including an explanation where no action is taken on a finding, need to be documented as to what was done and why [3]."

Requirement 5 – The two most recent compliance audit reports and documentation necessary to demonstrate the deficiencies have been corrected must be maintained. In practice, some companies choose to keep all audits for the life of the process and others discard prior audit reports and only keep the two most recent.

4 What does a PSM compliance audit tell you and how to use the results

This is where we get to the question in the title of this paper; **"Is 14 audit findings a good vital sign"?** Once a compliance audit has been completed, the focus then shifts to the number of audit findings per element, how to resolve the findings, and the priorities associated with the resolution. Does a fewer number of findings, less than 14, indicate a good vital sign? The answer to that question is maybe. In my opinion, the number of findings alone does not indicate much about the status and health of the PSM program.

Many variables can affect the number of audit findings that have little to do with the health of the program. Audit variables such as the number of auditors, skill / knowledge level of the auditor(s), number of audit days and the site resources available to answer questions; can all affect the number of findings.

The other factor, which can be more telling than the number of findings, is the type and severity of findings. For example, if the audit finding is that the facility has no MI program, this is more serious than having implemented a preventative maintenance (PM) program but failing to inspect one vessel out of ten. This in turn would be more serious than having inspected all ten vessels but having some administrative or recordkeeping errors, if the reports did not include all of the required items (i.e. missing the date or name of who completed the inspection, etc.).

How does one then determine the significance or severity of the finding? There are different approaches to determining the significance or severity of the audit findings. Both quantitative and qualitative methods can be applied to audit findings. In quantitative grading, a protocol or set list of questions is generally used and a value or number is assigned to each question. Normally a weighting is assigned where the questions judged to represent greater risk are worth more points than other questions. With this method, an overall score is normally assigned. In qualitative grading a set criterion for categorizing the findings and their relative importance must be established. A simple example would be a high-medium-low qualitative ranking. Each of these categories could also be assigned qualitative definitions similar to the severity and likelihood definitions used in a PHA.

To apply this to the MI example above; not having implemented a MI program would be scored high, failing to inspect one of ten vessels would be medium and missing required information on a completed inspection record would be low. So in theory having only fourteen findings could represent the worst case scenario of a facility having failed to develop or implement programs to comply with any of the fourteen PSM elements and having one finding for each element or having fully developed and implemented programs but just having some minor administrative or documentation issues.

For this reason, establishing and applying some form of grading scale to the audit findings is key. Scoring the findings will also help in developing the corrective action plan, as higher risk items should be assigned a priority for completion.

There are however potential downsides to both the quantitative and qualitative method. With quantitative scoring having one overall score can lead to comparisons being made between facilities or even at the same facility comparing the score of previous audits to the current one. With qualitative scoring, there is a tendency to try to compare the number of high-risk findings. The total score or number of findings by severity can be useful if you are comparing similar facilities, use a consistent audit team and apply a consistent scoring approach. This can be misleading if management tries to compare the scores of dissimilar facilities, the audit team was not consistent or the scoring criteria changes between audit cycles. If too much emphasis is placed on the score or number of findings by category or unreasonable expectations are assigned this can lead to pressure being applied to the auditor(s) to give more credit than is due in a quantitative system or not assign high risk in a qualitative system.

The author has experienced or observed the following scenarios over the last twenty plus years while working in the process industry and as a consultant. These scenarios have been organized as those causing a negative or a positive impact on the overall PSM program.

Negative: The following are viewed to be negative as they are examples of improper execution of the audit process and/or corrective action of the identified findings. As a result, the companies were not getting the full benefit of the audit process that is intended to identify gaps and take corrective actions that will improve the overall PSM program.

- A company tied the number of high-risk audit findings to the personnel performance metrics for some employees and a percentage of annual bonus for all employees. This led to an extreme amount of pressure being applied on both the audit participants and the auditors, for findings to not be categorized as high-risk. The majority of the discussion in daily and final close out meetings was around the ranking of the findings. Due to the focus on the risk category, participants did not use this time to understand what the findings were telling them about the health of the PSM program. They did not discuss what was contributing to the issues or how they could be corrected. As such, they missed the opportunity to get additional input from the auditors that could help to improve their PSM program.
- In another instance, the time allowed to close action items identified in an audit, was based on the severity of the finding. While it is a good idea to prioritize high-risk items, assigning an arbitrary time to completing them can cause issues. In this case, for all findings assigned as high the corrective action had to be closed within ninety days. In practice, this led to many of the corrective actions being written to address only a part of the problem so that it could be accomplished and closed within the ninety days. As an example, an audit finding identified that based on the relief system study some vessels did not have adequate protection installed. The corrective action assigned was to get a quote to design and install appropriate relief protection. The quote was obtained within the ninety days and the action item was closed. The issue was that three years later it was identified that the capital request to implement the project was not approved and the corrective measures were not implemented. The issue / risk still existed but there was no open action item to track the

actual installation of the required relief protection. The audit finding identified a clear risk or gap in the PSM program (a negative vital sign) but due to the focus on closing action items within the prescribed time, the corrective action was not managed in a way that reduced the risk.

- In an effort to try to learn from past audit findings and focus efforts to areas that showed the greatest gaps across most facilities, a company decided to change their scoring criteria in a quantitative grading system. Audit questions that typically had no findings written against them in prior audits were assigned less points and areas with larger gaps (more findings) were assigned more points. While this can be a good practice, it was poorly communicated to the sites and this led to issues with execution. The sites had developed employee performance goals tied to improving the audit score by a set percentage. Efforts were focused on improving areas where they had lost the most points on the previous audit and/or where they thought they could gain the most points on future audits. They did not know that corporate had changed the scoring criteria. As a result, improvement efforts were not focused on the areas that would have the greatest impact on the score under the new grading system. In practice with the change in grading, the overall score for most sites went down, even though the health of their overall programs had improved or at least stayed the same. Without understanding that the grading and therefore the expectations had changed, management was still trying to compare current audit scores to past results. Due to the poor communication and lack of understanding of the new scoring criteria, the health of the PSM program was incorrectly viewed to have gotten worse. Employees who had worked hard to implement the corrective action plan from the previous audit received negative criticism and performance ratings because of the perceived lack of improvement. This discouraged employee participation and hindered improvements to the PSM program.
- Compliance audits were viewed as a "have to do" activity as opposed to a value-added part of the PSM program. Audits were conducted by third party auditors and contracts awarded primarily based on costs. As the company had multiple facilities in different regions of the country different consulting companies, auditors and audit approaches were used with the quality of the audits varying greatly. As these audits were viewed as a check the box activity, little effort was given to understanding or correcting the issues that were identified. An incident at one facility led to an OSHA inspection with Notice of Violations (NOVs) and proposed penalties being issued. This prompted an OSHA inspection at another facility with the same result and so on. After multiple facilities in different states had been inspected and were facing proposed penalties the company entered into a Consent Decree with OSHA that included required changes to their audit process. In this case, prior to the OSHA penalties, the appropriate resources were not applied to conducting the audits or managing the resulting findings. The company was gaining little to no value from the audits because the audits often did not identify serious gaps which existed in their PSM programs and they did not properly manage the issues that were identified. The audits were not being used effectively to improve the overall health of the PSM program and reduce risks.

Positive: The following are viewed to be positive, as they are examples of proper execution of the audit process and/or corrective action of the identified findings. The companies invested the necessary resources and time to conduct the audits, revise the audit process when necessary, understand what the audit findings were telling them about the health of their programs and took appropriate corrective actions. As a result, the companies were getting the intended benefit of the audit process by identifying gaps and taking corrective actions that improved the overall PSM program thereby reducing risks.

- Past audit findings identified a lack of employee understanding and/or adherence to operating and emergency response procedures. Root cause analysis identified that a formal and consistent training program had not been implemented. Training was largely conducted via on-the-job training but there was no trainer qualification process or set requirements for what had to be communicated before an employee was considered to have completed training. A formal training procedure and plan was developed and implemented which included minimum requirements/qualifications to be a trainer, a train-the-trainer process, training curriculum by job position, written tests to gauge understanding and progress at set stages and a final review and approval process which included a panel interview and demonstration of knowledge before an employee was considered trained. After implementing this program, sites performed much better in this audit area, experienced fewer incidents and stated they observed production improvements related to fewer off-spec batches and less down time.
- Multiple audit cycles had identified issues related to tracking corrective actions to closure. It was identified that recommendations and associated corrective actions were coming from multiple sources or elements but there was no consistent way to track them to closure. The company had an electronic process for managing MOCs and action items were assigned and marked as closed within the system. They however had not established a process to maintain documentation of what had been done to consider the item closed. Therefore, actions like train all affected employees were marked closed but the site could not produce training records. In addition, actions identified in PHAs, incident investigations and prior audit reports were not being tracked in a centralized manner. PHA and audit findings were being tracked in spreadsheets specific to each PHA or prior audit and recommendations from incident investigations were only being tracked in each individual investigation report/folder. The company invested time and resources to implement an electronic tracking system where all recommendations and associated action items had to be entered into the system and assigned an ID number. For an action to be considered closed supporting documentation had to be attached within the system and verified by a second person. Having a centralized system allowed the company to have better visibility and understanding of the open issues at all facilities, prioritize resources between and within facilities to the higher risk items and establish metrics around closure of action items.
- Senior management recognized that their audit program was not delivering the expected results. The number of significant incidents and near misses experienced each year was not improving. Incident investigations identified issues such as lack of operating procedures or employee training as causal factors but these issues were not being identified in the audit process. In fact, the audits generally identified very few issues and rarely identified any as a high-risk. The majority of audits were conducted by internal resources often from the facility and process area being audited. The audit program used a checklist but this was largely based on document review and included limited or no infield verification of implementation. The decision was made to bring in outside resources to help redesign and implement a new audit process. In the new process, a combination of internal (company) and external (consultant) resources made up the audit team. The internal resources came

from other facilities or at least departments / process areas other than the one being audited. The audits continued to include a review of the relevant documentation and process safety information but also included inspection of the physical facilities, interviews with all levels of plant personnel and focused highly on infield observations of actual work practices. These new audits resulted in a significant increase in both the number of findings and those that were identified as high-risk. Despite some initial push back, senior leadership stayed with the process and were actively involved in determining and monitoring completion of corrective actions. Overtime they noted improvements in their overall PSM programs and a reduction in the number of significant incidents.

5 Conclusion

PSM Compliance audits have been required since 1995, despite being required for over 25 years the approaches used to conduct audits, manage the resulting recommendations and the value gained can vary greatly. An effective audit program is an essential element to the overall PSM program as it is the primary method to verify that systems to manage process safety are in place and functioning effectively. A poorly developed or executed audit program may not identify serious gaps that exist in the PSM program and therefore the opportunity to improve is missed. To have a successful audit program there must be support from all levels of the company and appropriate resources need to be assigned and allocated. A consistent audit approach must be developed and implemented that includes at a minimum, an audit method (protocol, experience based or combination), a grading system, auditor qualifications, a report identifying the findings and management of the resulting recommendations. Once a quality audit process is implemented, proper management of the resulting recommendations is arguably the most important aspect. The corrective actions need to address the cause, not just the symptom, to improve the overall health of the PSM program. This will likely require root cause analysis to determine why the systems are not achieving the expected results. The "why" must be understood in order to implement effective corrective actions. Corrective actions need to be prioritized and assigned reasonable expected due dates for completion. Facilities must maintain documentation necessary to demonstrate closure of the actions. The audit identifies the gaps; an effective corrective action plan must be developed and implemented within a timely manner for the benefit on an approved PSM program and reduced risks to employees and the community to be achieved.

As a vital sign or a health indicator of their PSM program, focusing just on the number of findings, establishing the wrong metrics, changing the metrics without proper communication or setting unrealistic deadlines for closing actions can have unintended consequences. Companies need to look beyond just the number of audit findings or score to understand what the audit is telling them about their processes and take appropriate actions to improve the PSM program. Fourteen high severity process related findings is a much worse vital sign than fourteen findings related to administrative or record keeping issues.

6 References

[1] CCPS. *Guidelines for Auditing Process Safety Management Systems, Second Edition.* Center for Chemical Process Safety, American Institute of Chemical Engineers, New York, NY, 2011.

[2] Occupational Safety and Health Administration, Process Safety Management, available at https://www.osha.gov/dsg/psm/index.html#:~:text=Unlike%20some%20of%20OSHA's%20standards,the%20unique%20conditions%20at%20their Accessed on February 9, 2021.

[3] Occupational Safety and Health Administration, Appendix C to §1910.119 -- Compliance Guidelines and Recommendations for Process Safety Management (Non-mandatory), available at https://www.osha.gov/pls/oshaweb/owadisp.show_document?p_id=9763&p_table=STANDARDS Accessed on February 10, 2021.

617884 Making Zero a Reality: Selecting & Measuring the "Right" Zero

Abstract:

Since the inception of the Process Industries (PI), leaders of these organizations have accepted some level of fires, explosions, releases to the environment and other events as the cost of doing business and *inevitable*. This cultural mindset continues to this day and is partially why, in spite of the many advances in process, occupational, and environmental safety made in the PI, it is still experiencing major accidents, fires & environmental releases.

Real safety performance and behaviors cannot be managed, much less produced, by corporate safety programs, many of which can be viewed as "bumper sticker" or "fake" safety, while active and latent hazards go unnoticed and/or unaddressed. The false sense of security that many PI safety programs encourage is evidenced by the continued occurrences of large-scale catastrophes. The dichotomy of having a "safety culture" versus such events is clearly evident, yet the PI continues to focus on and measure the same indices – all of which have little bearing on the cause of these events. *Zero* can be a goal, but it needs to be a realistic zero - a goal where no humans ever make an error is completely unrealistic. For many organizations the mistake isn't establishing Zero as a goal, the mistake is selecting the *wrong* Zero and then compounding the problem by failing to support the organization in order to make Zero a reality.

This paper examines the organizational policies and safety metrics of the PI with that of the single most reliable and safe organization in the world, which has achieved a safety record of ZERO incidents for its entire 65+ year history.

AUTHOR INDEX